TSUNAMIS IN THE EUROPEAN-MEDITERRANEAN REGION

TSUNAMIS IN THE EUROPEAN-MEDITERRANEAN REGION
From Historical Record to Risk Mitigation

GERASSIMOS PAPADOPOULOS
Institute of Geodynamics, National Observatory of Athens, Greece

ELSEVIER
Amsterdam • Boston • Heidelberg • London • New York • Oxford
Paris • San Diego • San Francisco • Singapore • Sydney • Tokyo

Elsevier
Radarweg 29, PO Box 211, 1000 AE Amsterdam, Netherlands
The Boulevard, Langford Lane, Kidlington, Oxford OX5 1GB, UK
225 Wyman Street, Waltham, MA 02451, USA

British Library Cataloguing-in-Publication Data
A catalogue record for this book is available from the British Library

Library of Congress Cataloging-in-Publication Data
A catalog record for this book is available from the Library of Congress

ISBN: 978-0-12-420224-5

For information on all Elsevier publications
visit our website at http://store.elsevier.com/

This book is dedicated to Ioanna with all my love.

CONTENTS

PREFACE

Tsunamis are one of the most important sea-related natural hazards. Practically speaking, coastal zones of all the oceans around the globe are exposed to tsunami hazard. This is also the case in the European and Mediterranean region where the historical tsunami record is the oldest on Earth. After the mega tsunami of December 26, 2004 in the Indian Ocean, which was very likely the most lethal in history, claiming around 220,000 human lives, the international community was mobilized aiming to develop actions for the tsunami risk mitigation. Writing this book is a contribution in this direction for it contains material which is useful not only to specialists but also to civil protection personnel and decision makers as well as to the general public.

After the introductory Chapter 1, which reviews fundamental global tsunami issues, including essays on the tragic tsunami events of 2004 in Indian Ocean and 2011 in Japan, the book covers all the aspects regarding the tsunami science, engineering, and risk mitigation in the European and Mediterranean region (EM).

Chapter 2 is an exhaustive overview of the historical evidence which is available for the tsunami documentation in the EM region. The geological record, both onshore and offshore, dated either in prehistoric or in historic times, has not been neglected. On the contrary, it is critically discussed as potential discriminator between tsunamis and other extreme sea waves, such as storm surges. The picture of the tsunami history is completed by archaeological observations. The instrumental documentation of the EM tsunamis in the last century or so is valuable for understanding important tsunami events, including the earthquake-generated tsunamis of 1908 in the Messina Strait (Italy), 1956 in the South Aegean (Greece), 1969 in Portugal, 1999 in Marmara Sea, and 2003 in North Algeria, as well as the landslide tsunamis of 1963, 2002, and 2014 in the west Corinth Gulf (Central Greece), Stromboli volcano (Italy), and western Norway, respectively.

Thanks to the long historical record, the tsunami impact in the EM region is also documented and analyzed in Chapter 3. An inventory of past tsunami impact has indicated that the main impact attributes are people killed and injured, damage to buildings, vessels, cultivated land, and to other property. In addition, of interest are also the various types of tsunami environmental impact, such as soil erosion, coastal geomorphological changes, boulder replacement and tsunami sediment deposition. On the basis of tsunami impact data, it was possible to assign tsunami intensity in the 12-point Papadopoulos and Imamura (2001) scale for the majority of the events listed in the historical record (Appendix). It has been found that historically the largest impact happened in the eastern Mediterranean Sea basin.

Chapter 4 focuses on the tsunami generation mechanisms. In the EM region, a variety of seismic and nonseismic tsunami source types have been recognized, such as seismic, volcanic, landslides. However, the generation mechanisms are precisely known only for a limited number of tsunamis. Even for instrumentally recorded events, such as the 1908 Messina and 1956 South Aegean tsunamis, the precise generation mechanism remain undetermined due to their high complexity which may involve several mechanism components, for example, seismic and landslide or volcanic and landslide, and so on. Criteria used to discriminate between the different source types include empirical relationships concerning the attenuation and the runup of tsunamis, as well as the slowness factor which characterizes the slip in the earthquake fault. The geographic zonation of tsunami sources is illustrated and some good examples of tsunami-generation mechanisms are analyzed. Rare types of tsunamis are also examined, such as meteotsunamis, tsunamis caused by asteroid impacts and from landslides in dams, as well as tsunami-like disturbances in lakes. Numerical simulations are a particularly valuable tool in the effort to understand the tsunami-generation mechanisms and source types. The relevant progress noted in the EM region is remarkable and internationally recognized. Therefore, this is particularly examined in Chapter 5.

The assessment of the tsunami hazard, vulnerability, and risk is a major issue since it is the basis to organize and apply land planning, crisis management, and other risk mitigation actions. Chapter 6 starts with a review of the glossary which is in use to express terms such as hazard, vulnerability, exposure, damage, intensity, risk, and the like. Then, the various methods and tools used for the assessment of the tsunami hazard, vulnerability, and risk in the EM region and beyond are examined. It has been found that significant progress has been made in the probabilistic and scenario-based assessment of the hazard and vulnerability although several sources or errors and uncertainties are involved. The tsunami risk assessment, however, is less developed. Therefore, more research effort should be put in this direction. On the other hand, the vulnerability and risk are strongly dependent on the degree of the community exposure to hazard. Of crucial importance are factors that control the time dependency of vulnerability and risk. All the various methods applied for the hazard assessment, qualitative, statistical, and probabilistic ones, agree that the most hazardous geotectonic structure in the EM region is the Hellenic Arc, in both the local (near-field) and remote (far-field) domains.

Chapter 7 provides an updated overview of the local, national, and regional tsunami early warning systems, either operational or prototype ones, that have been developed in the EM region particularly after 2004. Tsunami warning operations are supported by a variety of instrumental networks, algorithms, databases, communication technologies, and empirical tools. The most important regional system is the North East Atlantic and Mediterranean Tsunami Warning System, which is based on a number of national tsunami warning centers, for example, in France, Greece, Italy, Portugal, Turkey, and is coordinated by the ICG/NEAMTWS/IOC/UNESCO with the active collaboration of more than 30 country members. The activities of NEAMTWS are supported by the educational and training program of NEAMTIC. Of special value is also the JRC (European Commission) Tsunami Program which supports building up infrastructures, algorithms, databases, and training activities in emergency procedures. Good examples of local systems are the ones developed for early warning in the near-field domain against landslide-generated tsunamis in Stromboli Island (Italy), and against earthquake-generated tsunamis in Rhodes Island (Greece). More and more new emerging technologies appear, thus providing promises for the drastic improvement of the early warning services in the years to come. Of particular importance are also tsunami exercises, drills, communication tests, education, and training activities as well as other actions aiming to mitigate the tsunami risk in the EM region.

ACKNOWLEDGMENTS

My first involvement in the tsunami science dates back to 1983. Writing this book would not be possible without the research, educational and operational experience accumulated since then. Of particular value to me has been my participation in several relevant projects supported by the European Union, such as GITEC, GITEC-TWO, TRANSFER, SAFER, SCHEMA, SEAHELLARC, NEARTOWARN and ASTARTE as well as in national and bilateral projects supported mainly by the General Secretary for Research and Technology, Greece. Serving from various positions and with a variety of roles in the program of ICG/NEAMTWS/IOC/UNESCO, in the JRC Tsunami Program and in the national tsunami center of Greece, is highly appreciated.

In the last 30 years or so I have had the opportunity to collaborate with a large number of colleagues, students, and decision makers. I would like to extend to all of them my sincere thanks and deep appreciation for their constant spirit of collaboration, support, and friendship. Our collaboration in research projects, field expeditions, jointly writing papers, participating in conferences and various committees as well as exchanging ideas and experiences constitute valuable inputs for collecting material, organizing, and writing this book. The list is too long to mention all of them but certainly includes S. Tinti, A. Armigliato, R. Caputo, P. Gasparini, S. Lorito, A. Maramai, A. Michelini, D. Pantosti (Italy); A. Yalciner, N. Karanci, U. Kuran, O. Necmioglu, N. M. Ozel (Turkey); M.-A. Baptista, F. Carrilho, L. Matias, L.-V. Mendes (†) (Portugal); R. Dov, A. Salamon, A. Shapira (Israel); R. Guillande, H. Hebert, F. Lavigne, O. Lesne, F. Schindele (France); A. Dawson, T. Guymer (UK); F. Imamura, H. Matsumoto, K. Minoura, T. Nakamura, K. Satake, D. Sugawara, T. Takahashi (Japan); V. Gusiakov, B. Levin, L. I. Lobkovsky, R. Mazova, E. Pelinofsky, A. Rabinovich, E. Sasorova (Russia); D. Dominey-Howes (UK/Australia), S. Kortekaas (The Netherlands/Australia); B. Ranguelov (Bulgaria); A. Rudloff, J. Zschau (Germany); M. González, E. Gràcia (Spain); G. Georgiou (Cyprus); C. Ionescu, G. Marmureanu (Romania); C. B. Harbitz (Norway); A. Kijko, A Smit (S. Africa); E. Bernard, B. McAdoo, F. McCoy, V. Titov (USA); E. Argyris, Th. Dermentzopoulos, Ch. Koutitas, V. Lykousis, M. Papathoma, J. Papoulia, S. Pavlides, D. Sakellariou (Greece); A. Annunziato (JRC, EC) and O. Imperiali (DG-ECHO, EC).

I also thank warmly all my colleagues at the Institute of Geodynamics and the Hellenic National Tsunami Warning Center, National Observatory of Athens (Greece), particularly M. Charalambakis, E. Daskalaki, A. Fokaefs, A. Ganas, V. Karastathis, N. Liadopoulou, G. Minadakis, T. Novikova, K. Orfanogiannaki, A. Plessa, G. Stavrakakis (†), A. Tselentis, to mention a few. Thanks to their encouragement, support, and collaboration that made it possible not only to promote the tsunami research at our institute but also to build up and run the national tsunami-warning center.

My close collaboration with the Tsunami Unit of IOC/UNESCO goes back to the 1990s. I am grateful to all the people that served and still serve with IOC for their constant support, collaboration, and friendship.

G.A. Papadopoulos
August 2015

INTRODUCTION

A tsunami is a series of sea waves with long period and long wavelength generated by an abrupt deformation of the seafloor or by other sudden disturbance of the sea water level. The energy of vertical movement of such a disturbance is transferred to the water mass and causes a sea level change at the source region. Underwater and/or coastal earthquakes, volcanic eruptions, as well as landslides are natural processes that can generate a tsunami. However, strong earthquakes remain the most frequent cause of tsunamis. Large meteorites that may impact the ocean should not be ruled out as possible agents of tsunami generation. One should not neglect anthropogenic actions that may result in tsunami production, for example, submarine nuclear bomb testing. Tsunami waves propagate outward from the generating area in all directions, the main direction of energy propagation being controlled by the dimensions and orientation of the causative source. During its propagation in deep water, the tsunami proceeds as a series of ordinary gravity waves with a speed depending on the water depth. In the near-shore domain, a large amount of energy is carried by both the amplified water level and strong currents. Hence, tsunamis cause scouring, erosion, deposition, slope failures as well as damage or even destruction in coastal communities, marine structures and other facilities, cultivated land, and natural environment.

In the European-Mediterranean (EM) region, tsunami generation has been documented with frequency that varies from one area to another. In the Mediterranean region, the seismic activity is high due to active geodynamic processes such as the convergence of the Eurasian and African lithospheric plates (Argus et al., 1989; De Mets et al., 2010). Subduction of oceanic crust and/or collision takes place along active orogenic belts, namely from west to east, the Gibraltar Arc, the Calabrian Arc, the Hellenic Arc, and the Cyprus Arc. Ongoing motion along transform boundaries of adjacent plates, such as the Arabian plate and smaller crustal blocks, such as the Anatolian "micro-plate," add more complexity to the active Mediterranean geodynamics and the resulting geological processes (see also in Mascle and Mascle, 2012). Instead, the regions of the North East Atlantic Ocean, the North Sea and the Baltic Sea are characterized by low seismicity, with the exception of the area offshore South West Iberia. The Black Sea is also of low seismicity. Therefore, the tsunami activity is much higher in the Mediterranean, including the area offshore of South West Iberia, as compared to the other European sea regions. Although the frequency of tsunamis in the Mediterranean is much less than the frequency of strong earthquakes, tsunamis threaten seriously the communities along the coastal zones of the Mediterranean basin (e.g., CIESM, 2011). This is also the case for the rest of the European coastal zones.

Tsunami generation in the EM region is dependent on several geodynamic processes taking place in submarine or in coastal environments, which mainly include earthquake activity, several mechanisms associated with volcanic eruptions, and landslide processes. Tsunami activity from prehistorical times up to the present has been documented in the EM region from a variety of information sources. In fact, geological evidence, mainly onshore and more rarely offshore tsunami sediment deposits, geomorphological features, historical documentary sources, archeological findings, as well as instrumental data and observations collected during post-event field surveys, have provided a long record of tsunami events (Papadopoulos et al., 2014a). Only a few of those events were basin-wide; others, however, were either regional or only local tsunamis. In addition, in the EM region it has been reported the presence of the so-called meteotsunamis (Monserrat et al., 2006) that are tsunami-like sea waves attributed to atmospheric changes rather than to seismic and other geodynamic processes.

With the aim to organize actions toward the tsunami risk mitigation, there is need to have a good knowledge of the past tsunami activity, to identify potential tsunami sources and better understand the generation mechanisms particularly in the frame of complex geodynamic setting such as the one which, for example, characterizes the Mediterranean region. Another important issue is to characterize tsunami sources, in other words to improve our capabilities to discriminate between seismic and aseismic mechanisms. As a matter of fact, for many historical and recent tsunami events the causative sources and generation mechanisms still remain unidentified.

After the devastating Indian Ocean tsunami that generated in Sumatra on the December 26, 2004, a systematic effort was put forward for building up the North East Atlantic and Mediterranean Tsunami Warning System (NEAMTWS), beginning in November 2005. The country members of the Intergovernmental Oceanographic Commission (IOC) of UNESCO, working together under the umbrella of IOC, have established the NEAMTWS which today functions in an interim operational status based on the synergy of a number of national tsunami warning centers. In parallel to this, other actions aiming at tsunami risk mitigation, such as tsunami alerting mechanisms, drills and educational/training activities, have been developed by several EM countries and international organizations, including the Joint Research Center, the DG ECHO and other directorates of the European Commission, and the NEAMTIC which is the Tsunami Information Center of the NEAMTWS.

This book is the first to focus on the tsunamis of the EM region. All aspects of tsunamis are included, namely, documentation and impact of tsunami events from prehistory up to the present; examination of the tsunami sources and generation mechanisms; evaluation of lessons learned from the tsunami numerical modeling; review of the methods in use for the tsunami hazard, vulnerability and risk assessment; up-to-date description of the existing early warning systems and of other actions aiming to risk mitigation, such as education, preparedness and crisis management. A catalogue of important tsunamis reported in the entire EM region is also included. However, in order to get a comparative view with the developments in tsunami research and risk mitigation in other parts of the Earth, the first chapter of the book is devoted to a short review of tsunamis around the globe.

CHAPTER 1

Tsunamis in the Global Ocean

1.1 TSUNAMIS AND MEGATSUNAMIS

According to the definition of Van Dorn (1968), *tsunami is the Japanese name for the gravity wave system formed in the sea following any large-scale, short-duration disturbance of the free surface.* The occurrence of tsunami events has been reported in all parts of the global ocean. The documentation of tsunami events that occurred in the preinstrumental period depends on the availability of a variety of records based on geological and archaeological observations and on historical documentary sources. For this reason the tsunami reporting period varies from one side of the global ocean to the other. For example, in Greece and other coastal sites of the east Mediterranean basin one may find the oldest historical tsunami documentation thanks to the availability of relevant historical sources and archaeological observations (Papadopoulos et al., 2014a).

Modern tsunamis are mainly instrumentally documented, which today includes records in tide gauges at the shorelines and in pressure tsunameters on the sea floor. In the last two decades or so efforts have been made toward detecting tsunamis propagating in the open ocean by satellite altimetry methods (see a review in Levin and Nosov, 2009). Of particular value are field data and eyewitnesses accounts collected during post-event field surveys as well as pictures and videos taken in coastal spots hit by tsunamis. In the last 20 years or so several large, disastrous tsunamis occurred in both the Pacific and Indian Oceans. Two of them, the 2004 Sumatra tsunami and the 2011 Japan tsunami, were among the largest ever reported globally, not only because of the physical dimension of their size but also due to their highly catastrophic consequences. In fact, they were ocean-wide tsunamis propagating across the Indian Ocean first and the Pacific Ocean second, while in both cases the measured maximum wave runup exceeded 30 m.

Such very large tsunamis are termed "megatsunamis", although no standard definition has been given to define them (Goff et al., 2014). One may recognize that the historical record contains many examples of similar basin-wide tsunami waves associated with high run-up, such as the Chile seismic tsunami of May 22, 1960 in the Pacific Ocean and the Krakatoa volcanic tsunami of August 27, 1883 in the Indian Ocean. Table 1.1 lists some

Tsunamis in the European-Mediterranean Region. http://dx.doi.org/10.1016/B978-0-12-420224-5.00001-6

Table 1.1 List of some significant tsunamis reported worldwide in the last 150 years or so

Date	Source area	Generation mechanism	H_{max} (m)	Distant areas affected
August 13, 1868	Arica (Chile)	EQ	21.0	Peru, Japan, Hawaii, New Zealand, Australia, Fiji, USA
May 10, 1877	Arica (Chile)	EQ	24.0	Peru, California, Hawaii, New Zealand, Australia
August 27, 1883	Krakatau Volcano (Indonesia)	VE	35.0	All around Indian Ocean
August 15, 1918	Celebes Sea	EQ	12.0	Indonesia, Philippines
September 7, 1918	S. Kuril Islands	EQ	12.0	Japan, Hawaii
November 11, 1922	Atacama (Chile)	EQ	12.0	Colombia, Hawaii, Japan, Samoa, New Zealand
March 2, 1933	Sanriku (NE Japan)	EQ	29.0	Hawaii
April 1, 1946	Unimak Island (Alaska)		35.0	Hawaii, Peru, California, Samoa, Chile
November 4, 1952	Kamchatka Peninsula	EQ	20.0	Hawaii, Sanriku (Japan)
March 9, 1957	Aleutian Islands	EQ	16.2	Hokkaido (Japan), Hawaii, California, El Salvador
May 22, 1960	S. Chile	EQ	25.0	South-Central-North America, Hawaii, Japan, Marquesas Islands, Samoa, Kuril Islands, Taiwan, Fiji, New Zealand, Australia
March 28, 1964	Alaska	EQ	67.0	US west coast, Canada, Hawaii, Japan
August 19, 1977	S. Sumbawa	EQ	15.0	Australia
June 2, 1994	Java	EQ	13.9	Australia
December 26, 2004	Sumatra	EQ	30.0	All around Indian Ocean, Australia
February 27, 2010	Maule (Chile)	EQ	14.0	California, Tohoku (NE Japan)
March 11, 2011	Tohoku (NE Japan)	EQ	40	Japanese west coast, Hawaii, California

Tsunamis in the European-Mediterranean region are not included here since a more detailed list can be found in Chapter 2. Key: EQ, earthquake; VE, volcanic eruption; H_{max}, maximum wave height.
Sources: Lockridge (1988), Intergovernmental Oceanographic Commission (1999) (see also summary in Bryant, 2008).

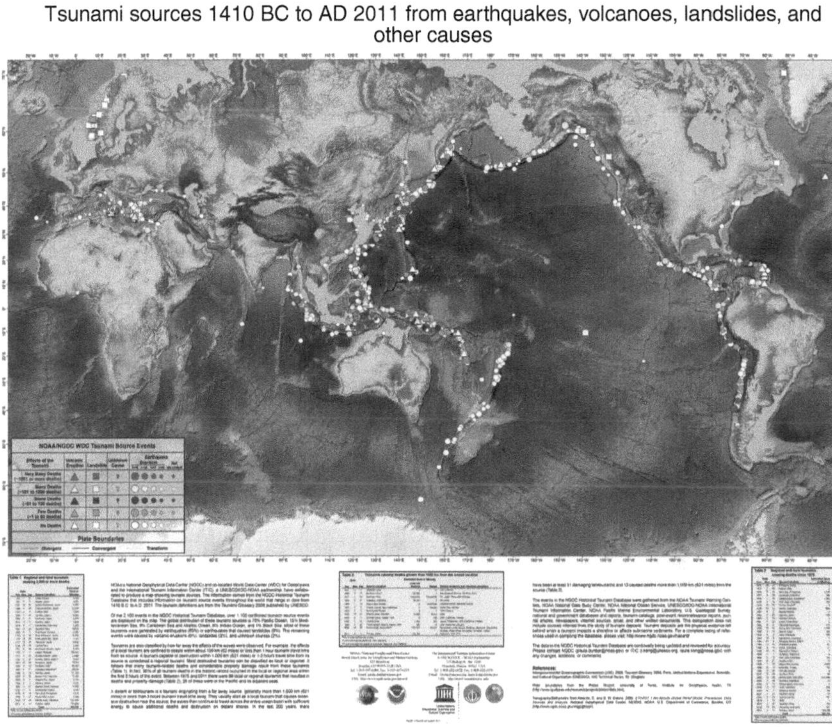

Figure 1.1 ***Global map of known tsunami sources from the antiquity up to 2010.*** *(Source: NOAA/USA, http://www.ngdc.noaa.gov/hazard/tsu.shtml.)*

of the most significant tsunamis in the Pacific and Indian oceans during the last 150 years or so. However, analogous historical cases can be found in the European and Mediterranean region as well. For example, it is widely accepted that the concept of megatsunami could apply to the seismic tsunamis of AD July 21, 365 and August 8, 1303 in Crete Island (Hellenic Arc) and of November 1, 1755 in South West Iberia, Atlantic Ocean. One may not exclude the large tsunami, which is documented by geological and archaeological evidence to have occurred around the end of the seventeenth century BC after the colossal Plinian-type eruption of the volcano of Thera (Santorini) in the South Aegean Sea. Figure 1.1 shows a global map of known tsunami sources from the antiquity until the present day.

After the devastating events of 2004 in Sumatra and 2011 in Japan the term megatsunami became quite popular. However, as Goff et al. (2014) noted no unambiguous and widely accepted definition currently exists for that term. These authors proposed a stricter definition for megatsunami that is based solely on initial wave height/amplitude at source of 100 m/50 m, respectively.

1.2 TSUNAMI RECORD AND IMPACT

The tsunami documentation is supported by databases which are usually organized by seismological or meteorological institutes around the globe. However, such databases are not organized in a standardized way and often suffer from a lack of updating. A good example of tsunami documentation is the New European Tsunami Catalogue, which was initially organized by a group of scientists working together in the frame of the tsunami research projects GITEC and GITEC-TWO supported by the European Commission during the 1990s. Later, with the next pan-European EU-FP6 project TRANSFER (2006–2009) the database was completed and updated, while refinements are expected from the ongoing pan-European EU-FP7 project ASTARTE (2013–2016). Table 1.1 lists some of the most important tsunamis reported historically around the globe.

The impact of tsunami events on communities as well as on the natural environment is controlled by several factors, some of them favoring and others disfavoring damage and destruction. Physical factors dominate, while other factors are dependent on the built environment and characteristics of coastal communities. A critical physical factor is the initial tsunami size at its source, which is due to the generation mechanism. However, the characteristics of the tsunami (e.g., wave amplitude, velocity, period) do not depend only on the size and type of source and the generation mechanism but also on the ocean bathymetry as the tsunami propagates outward from the source. The height and other tsunami wave characteristics at a particular coastal site are drastically influenced by the bathymetry in the shallow-water domain. Finally, the inundation (flooding) of the tsunami in coastal areas is determined by the features of the coastal environment, such as the presence of forest or other vegetation or sand dunes and the overall coastal geomorphology.

The characteristics of the coastal communities, such as the number and type of buildings, vessels, cultivated land, land use/land cover, road network, fisheries, and infrastructures, combined with the degree of vulnerability and exposure to the tsunami threat are crucial factors that determine the final impact in a particular coastal zone. One may understand that it is not an easy task to estimate the potential impact of future tsunami events. This problem is examined when the tsunami hazard, vulnerability, and risk assessment issues are more extensively discussed (Chapter 6). Examples of tsunami impact referring to the megatsunamis of 2004 in the Indian Ocean and of 2011 in Japan are presented in later sections of this chapter, while the impact of some large historical tsunamis in the European-Mediterranean region is presented in Chapter 3.

1.3 TSUNAMI SOURCES AND GENERATION MECHANISMS

Earthquake activity is the most frequent cause of tsunamis. During the occurrence of a strong earthquake, the coseismic fault displacement at the sea bottom pushes upward the seawater column. Then, the displaced water mass collapses due to gravity and the tsunami is generated as a gravity wave that propagates outward from its source. The fault displacement and the dimensions of the fault segment involved in the tsunami generation determine the tsunami size at the source. The complexity of the seismic rupture is also important. In fact, the initial size of the tsunami wave may vary with the homogeneous or heterogeneous rupture along the fault plane (Geist and Dmowska, 1999) as well as with the friction pattern during the rupture (Bilek and Lay, 1999).

However, the concurrence of several geophysical factors is needed for the generation of a tsunami, such as shallow earthquake focus (focal depth less than 100 km) and large magnitude (usually no less than about 6.5) of the causative earthquake. Also, the focal mechanism plays an important role. In fact, the dip-slip type of seismic faulting (normal or reverse) favors tsunami generation since it involves a significant vertical component in the coseismic fault displacement. The largest earthquakes and tsunamis occur in active zones of lithospheric subduction where the earthquake focal mechanism is predominantly reverse faulting (Figure 1.2). Strike-slip ruptures, where the horizontal component of fault motion dominates, does not exclude but disfavor tsunami generation. For example, in the Aegean area, the large earthquake (magnitude $M = 7.5$) that ruptured in the area of Cyclades Islands, South Aegean Sea, on July 9, 1956 was associated with submarine

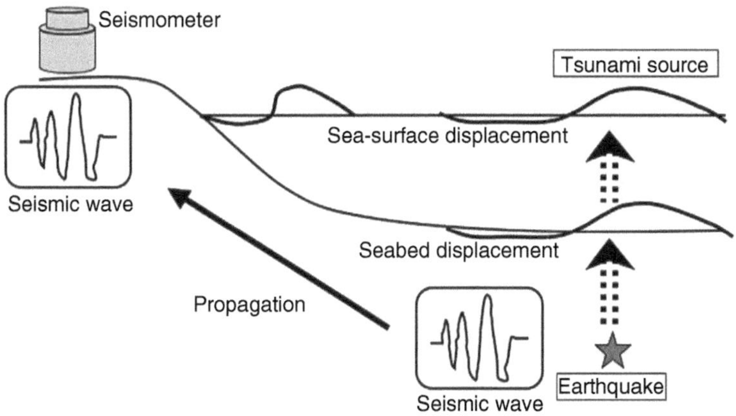

Figure 1.2 ***Tsunami generation due to coseismic seabed dislocation (Takahashi, 2006).***

normal faulting, which is believed to have favored the production of a large tsunami with amplitude in the near-field domain of up to about 15 m (Papadopoulos and Pavlides, 1992; Beisel et al., 2009; Okal et al., 2009; see more details in Chapter 4). On the other hand, in the North Aegean Sea where the strike-slip component dominates in the seismic ruptures, none of the large earthquakes of February 20, 1968 ($M = 7.1$), December 19, 1981 ($M = 7.2$), January 18, 1982 ($M = 7.0$), and May 24, 2014 ($M = 6.9$) caused tsunamis.

Several mechanisms can be recognized in the tsunami generation during volcanic eruptions. They may include volcanic earthquakes, caldera or cone collapse, pyroclastic flows, and more (Latter, 1981). For the large Minoan tsunami caused by the Late Bronze Age eruption of Thera (Santorini) volcano by the end of the seventeenth century BC, two main mechanisms were proposed and tested by numerical simulations: caldera forming collapse of the volcanic cone and massive pyroclastic flows rolling down the volcanic cone seaward (Minoura et al., 2000; Pareschi et al., 2006a; Novikova et al., 2011). The second mechanism can be considered as a particular type of volcanic landslide. A more conventional case of volcanic landslide is the one where the volcanic activity triggers the landslide of unstable masses of volcanic and/or other rocks. This happened with the volcanic activity in Stromboli volcano, Aeolian Islands, Italy, on December 30, 2002. The local tsunami produced had a height of ca. 9 m and caused some damage only to outdoor and indoor property (Tinti et al., 2005c). However, coastal or submarine landslide in nonvolcanic areas is also a well-known mechanism for tsunami generation. Such landslides may be due to seismic activity or only to the gravity force. It is assumed that in all these mechanisms the seawater is abruptly displaced and then collapses creating gravity sea wave.

In the European–Mediterranean region, all the earlier discussed different tsunami generation mechanisms have been recognized. Therefore, it is of value to introduce some terminology with the aim of better describing and distinguishing between such mechanisms. Here I follow the suggestion I made several years ago as regards this particular issue (Papadopoulos, 1993a) (Table 1.2). The term *seismic tsunami* refers to tsunami generation that results from coseismic fault displacement of the sea floor. An earthquake that generates tsunamis with this mechanism is called a *tsunamigenic earthquake* (Figure 1.3). However, this term should not be confused with the term *tsunami earthquake* in the terminology introduced by Kanamori (1972) to characterize the 1992 Nicaragua earthquake source and other earthquakes whose tsunamis were disproportionately large with respect to

Table 1.2 Classification of tsunamis according to their generation mechanism

Seismic tsunamis: produced by tsunamigenic earthquakes (mechanism: coseismic fault displacement)	
Nonseismic tsunamis: without direct involvement of earthquake activity	
Aseismic tsunamis: due to landslides, volcanic activity, or other causes without the involvement of earthquake activity	*Pseudoseismic tsunamis*: due to landslides or volcano collapse due to earthquake activity acting only as a triggering factor

Source: After Papadopoulos (1993a).

their size measured either by seismic moment or by magnitude. Any other mechanism of tsunami generation is *nonseismic*. This term, however, may refer to two alternatives. The first includes an earthquake as only a triggering factor, for example, of a coastal or submarine landslide (Figure 1.4) or of the collapse of a submarine volcano because of the earth shaking. Such a mechanism of tsunami generation is called *pseudoseismic* since no coseismic fault displacement is involved. However, when the landslide occurs without any seismic triggering then the mechanism of tsunami generation is purely *aseismic*.

Large meteorites that may impact the ocean should not be ruled out as possible agents of tsunami generation. It is suggested that this happened with the very large impact-induced tsunami that occurred at Chicxulub, Mexico, at the Cretaceous-Tertiary boundary around 65 million years ago and possibly was associated with the extinction of the dinosaurs. However, such impacts are quite rare. Also, one should not neglect anthropogenic

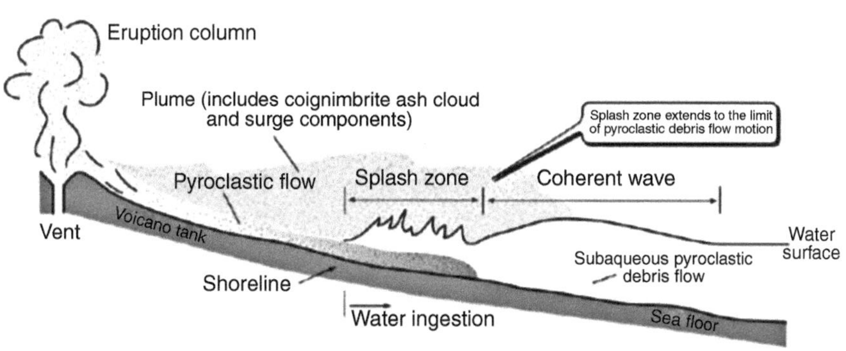

Figure 1.3 ***Schematic diagram of tsunamigenic mechanism from pyroclastic flow at the slope of a volcanic cone with plume entering the water (Novikova et al., 2011, redrafted from Watts and Waythomas, 2003 and Walder et al., 2003).***

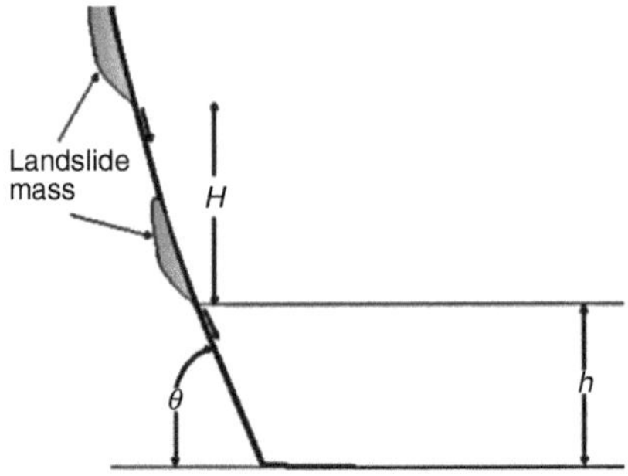

Figure 1.4 ***Tsunamigenic landslide mass released either at the shore or at a height, H, above water level accelerates down a steep slope and only decelerates when it reaches the bottom at depth h (Novikova et al., 2011, redrafted from Watts and Waythomas, 2003 and Walder et al., 2003).*** **For the caldera collapse tsunamigenic mechanism of the Santorini (Thera) LBA eruption, Novikova et al. (2011) considered a dynamic series of landslides.**

actions that may result in tsunami production, for example, submarine nuclear bomb testing.

1.4 TSUNAMI PHYSICAL PROPERTIES: A FEW ELEMENTS

Under the assumption that depth is small compared to a horizontal length scale, there are three regions of approximation for the long-wave theory: (a) linear equations, (b) finite-amplitude equations, (c) Boussinesq equations (Murty, 1977; Levin and Nosov, 2009). Three characteristic lengths determine which equation is most appropriate: water depth, D, wave length, λ, and wave amplitude, η. It has been shown that for very long waves, which is the case of tsunamis in the deep ocean, the speed of the wave is mainly controlled by the water depth as a first approximation:

$$C = \sqrt{gD} \tag{1.1}$$

where g is the acceleration due to gravity. In the theory of tsunami propagation and relevant numerical modeling applications, both linear and nonlinear wave equations have been utilized. For tsunamis traveling in the near-shore domain, that is, over the continental shelf, it is more appropriate to use Boussinesq equations which are of intermediate type. In tsunami wave propagation, *frequency dispersion* or phase dispersion or more simply just dispersion, as well as *amplitude dispersion* or nonlinear effects, occur. The meaning of frequency dispersion is that wave components of different frequencies propagate with different velocities. On the other hand, when greater values of surface elevations propagate with greater velocities then amplitude dispersion takes place. Steepening of the wave occurs because of amplitude dispersion.

In the near-field domain of tsunami source, which as a rule is characterized by deep water, linear wave theory is adequate to describe the tsunami propagation given that practically only phase dispersion applies. As the tsunami propagates near or over the continental shelf, both phase and amplitude dispersions become important and the concept of a *solitary wave* becomes appropriate. In the very shallow water domain, however, and particularly in closed bays, harbors, and inlets, the amplitude dispersion dominates which results in a significant increase of the slope of the wave which usually is considered as a *train wave.*

In the deep ocean the wave speed may approach the speed of sound, which is 1200 m/s or about 4300 km/h. As the wave travels over shallower water, it slows down. As the wave comes very close to the coast, the tsunami speed reduces to about 50–60 km/h. In deep water the wave amplitude remains small, around 1 m. However, in the coastal zone the wave heights may exceed 30 m or 40 m. For deep water waves $\left(\frac{D}{\lambda}>\frac{1}{2}\right)$ the water particle orbits are closed circles with the radius decreasing with depth; water pressure decays too with depth. For shallow-water waves $\left(\frac{D}{\lambda}<\frac{1}{2}\right)$ the phase velocity as expressed in equation (1.1) is independent of the wave length or the wave period, and, therefore, the waves are nondispersive. The particle orbits are ellipses with the major axis horizontal. As the length of the major axis remains constant at all levels and the minor axis decreases with depth, close to the bottom the ellipse almost becomes a straight line (Murty, 1977). The period of tsunami waves in most cases ranges between 5 min and 90 min.

1.5 TSUNAMI QUANTIFICATION

A tsunami can be considered as a particular case of seismic wave. In fact, Okal (1988) showed that source depth and focal geometry play only a limited role in controlling the amplitude of the tsunami, and that more important are the effects of directivity due to rupture propagation along the fault and the possibility of enhanced tsunami excitation in material with weaker elastic properties, such as sedimentary layers. In view of this, issues related to tsunami quantification could be approached in analogy to seismology. However, in tsunami science many developments were delayed by some decades with respect to progress achieved in seismology. This is because tsunamis are infrequent events compared to earthquakes, and this in turn caused delays in developing effective observation systems and measurement practices.

1.5.1 Earlier Scales

Efforts toward quantifying earthquakes in terms of macroseismic intensity began in the mid-nineteenth century. After the initial introduction of 6-grade and 10-grade scales, such as the Rossi-Forel one in Italy in 1883, and 7-grade scales in Japan in 1900, efforts concluded with the gradual introduction of the 12-grade Mercalli–Cancani–Sieberg scale in 1912, the Modified-Mercalli scale in 1931 and 1956, the Medvedev–Sponhuer–Karnik scale in 1964 (see review in Coburn and Spence, 1992) and finally the European Macroseismic Scale in 1998. The concept of earthquake magnitude was introduced by Charles Richter (1935) to measure the size of earthquakes on a physically based scale, the so-called local magnitude scale. As a next step, the introduction of the concept of seismic moment by Aki (1966) opened the way for the establishment of the moment-magnitude scale (Kanamori, 1977). The earthquake magnitude is an objective physical parameter that measures either energy radiated by, or moment released in, the earthquake source and does not reflect macroseismic effects. On the contrary, the earthquake or seismic intensity is an estimate of the event impact. An earthquake is characterized by different intensities in different observation points of the affected area.

In regard to the quantification of tsunami size, an early effort can be found in the pioneering work of Sieberg (1923, 1927) who defined the first 6-grade tsunami intensity scale based on the tsunami effects. However, the tsunami quantification remained as a puzzling aspect in tsunami science since the scales proposed later to measure tsunami size often were either difficult to apply or confusing as for the quantity they represented.

Ambraseys (1962) published a modified version of Sieberg's 6-grade scale known as Sieberg–Ambraseys tsunami intensity scale. In the Japanese tsunami literature one may find a long tradition of effort in tsunami quantification. Imamura (1942, 1949) introduced and Iida (1956, 1970) and Iida et al. (1967) developed further the concept of tsunami magnitude, m, defined as

$$m = \log_2 H_{max} \tag{1.2}$$

where H_{max} is the maximum tsunami wave height (in meters) observed in the coast or measured in the tide gauges. Practically, the so-called Imamura–Iida scale is a 6-point scale ranging from −1 to 4 giving the impression of an intensity rather than a magnitude scale. However, *m* does not estimate effects but it measures by definition H_{max}, which is a physical quantity. In this sense it may represent magnitude in a primitive way since it does not calibrate the wave height with distance. In his attempt to improve Imamura–Iida's definition, Soloviev (1970) proposed to define tsunami intensity, i_S, by

$$i_s = \log_2 \sqrt{2}\ (H) \tag{1.3}$$

H (in meters) is the mean tsunami height in the coast. However, this is still a primitive magnitude scale, since it is also based on the physical quantity *H*. Tsunami magnitude M_t (Abe, 1979, 1985, 1989) or *m* (Hatori, 1986) was defined by the general form

$$M_t = a \log_{10} H + b \log \Delta + D \tag{1.4}$$

where *H* is the maximum single (crest or trough) amplitude of the tsunami waves (in meters) measured by tide gauges, Δ is the epicentral distance (in kilometers) of the tide station along the shortest oceanic path (in kilometers), and *a, b, D* are constants. Expression (1.4) is similar to the Prague formula (Vaněk et al., 1962) used since the 1960s for the measurement of the surface-wave earthquake magnitude. A different approach for the calculation of the tsunami magnitude was introduced by Murty and Loomis (1980). Their tsunami magnitude, ML, was defined by

$$\mathrm{ML} = 2\left(\log_2 E - 19\right) \tag{1.5}$$

E is the tsunami potential energy (in ergs). Definition of ML is in close analogy to Kanamori's (1977) definition of moment magnitude, M_w, as well as to the mantle magnitude, M_m, introduced by Okal and Talandier (1988), where M_0 is seismic moment:

$$M_w = \frac{2}{3}\left(\log_{10} M_0 - 16.1\right) \tag{1.6}$$

$$M_m = \log M_0 - 20 \tag{1.7}$$

A particular scale measuring tsunami size is the one proposed by Shuto (1993) who considered it as an intensity scale:

$$i = \log_2 H \tag{1.8}$$

Table 1.3 Empirical correlation between the 12 intensity domains, *K*, of the Papadopoulos and Imamura (2001) new tsunami intensity scale and the quantities *H* and *i* introduced in formula (1.8) by Shuto (1993)

K	*H* (meters)	*i*
I–V	<1.0	0
VI	2.0	1
VII–VIII	4.0	2
IX–X	8.0	3
XI	16.0	4
XII	32.0	5

H is the local tsunami height (in meter) and *i* is a measure of the tsunami size.

However, it is still a magnitude scale by definition; *H* is the local tsunami height (in meters). However, in order to use it as an intensity scale for tsunami damage description, Shuto (1993) proposed to define *H* according to its possible impact. A 6-point empirical classification of tsunami size ranging from 0 to 5 was tabulated for the description of the expected impact as a function of *H* (Table 1.3).

The need to replace 6-point tsunami intensity scales with modern, detailed 12-point scales in analogy to earthquake intensity scales was noted by Papadopoulos and Imamura (2001). They also underlined that the proposed tsunami magnitude scales, even the most sophisticated ones, have been of very limited practical use since they need either better calibration of formulas based on more wave height data or significant improvement in the tsunami source energy calculation. As a matter of fact, tsunami magnitude scales that are based on measurements of tsunami wave heights at coastlines are very sensitive to local effects, such as coastal topography, near-shore bathymetry, refraction, diffraction, and resonance. However, better calibration of formulas, based on more tide gauges as well as wave heights measured in the field, may drastically improve the applicability of such scales for the tsunami magnitude determination in the future. On the other hand, the Murty–Loomis tsunami magnitude, which is directly based on the total tsunami energy, *E*, at the source, provides a wider magnitude range but is not easily applicable because of serious difficulties involved in the calculation of energy, *E*. Better estimates of tsunami energy in the future, based on tsunameters record, certainly will result in the magnitude determination of a more and more increasing number of tsunamis.

1.5.2 The New Papadopoulos-Imamura Intensity Scale

Having these problems in mind, Papadopoulos and Imamura (2001) introduced a new 12-grade tsunami intensity scale as a realistic proxy of tsunami size (see also Papadopoulos, 2003a). This new scale is based on three principles: (a) independency from any physical parameter; (b) sensitivity, that is incorporation of an adequate number of divisions (or points) in order to describe even small differences in tsunami effects; and (c) detailed description of each intensity division by taking into account all possible tsunami impacts on the human and natural environment, the vulnerability of buildings and other engineered structures. The new tsunami intensity scale incorporates 12 divisions and in this sense is consistent with several 12-point seismic intensity scales established and extensively used in Europe and North America in about the last 100 years or so. The new scale is arranged according to (a) the effects on humans, (b) the effects on objects, including vessels of variable size, as well as on the natural environment, and (c) damage to buildings and other engineered structures.

A correlation between intensity domains of the 12-point scale and the intensity size proposed by Shuto (1993) is shown in Table 1.3, while a full description of the 12-point tsunami intensity scale can be found in Box 1.1.

BOX 1.1 The new 12-point tsunami intensity scale by Papadopoulos and Imamura (2001) (see also Papadopoulos, 2003a).

I Not felt

- Not felt even under the most favorable circumstances.
- No effect. No damage.

II Scarcely felt

- Felt by few people on board small vessels. Not observed at the coast.
- No effect. No damage.

III Weak

- Felt by most people on board small vessels. Observed by few people at the coast.
- No effect. No damage.

IV Largely observed

- Felt by all on board in small vessels and by few people on board large vessels. Observed by most people at the coast. Few small vessels move slightly onshore.
- No damage.

V Strong

- Felt by all on board large vessels and observed by all at the coast. A few people are frightened and run to higher ground.
- Many small vessels move strongly onshore, few of them crash into each other or overturn. Traces of sand layer are left behind in grounds of favorable conditions. Limited flooding of cultivated land.
- Limited flooding of outdoor facilities (e.g., gardens) of near-shore structures.

VI Slightly damaging

- Many people are frightened and run to higher ground.
- Most small vessels move violently onshore, or crash strongly into each other, or overturn.
- Damage and flooding in a few wooden structures. Most masonry buildings withstand.

VII Damaging

- Most people are frightened and try to run to higher ground.
- Many small vessels damaged. Few large vessels oscillate violently. Objects of variable size and stability overturn and drift. Sand layer and accumulations of pebbles are left behind. Few aquaculture rafts washed away.
- Many wooden structures damaged; some are demolished or washed away.
- Damage of grade 1 and flooding in a few masonry buildings.

VIII Heavily damaging

- All people escape to higher ground, a few are washed away.
- Most of the small vessels are damaged, many are washed away. A few large vessels are moved ashore or crash into each other. Large objects are drifted away. Erosion and littering at the beach. Extensive flooding. Slight damage in tsunami control forest, stop drifts.
- Many aquaculture rafts washed away; a few are partially damaged.
- Most wooden structures are washed away or demolished. Damage of grade 2 in a few masonry buildings.
- Most Reinforced Concrete (RC) buildings sustain damage, in a few damage of grade 1 and flooding is observed.

IX Destructive

- Many people are washed away.
- Most small vessels are destroyed or washed away. Many large vessels are moved violently ashore, some are destroyed. Extensive erosion and littering of the beach. Local ground subsidence. Partial destruction in tsunami control forest, stop drifts. Most aquaculture rafts washed away, many partially damaged.
- Damage of grade 3 in many masonry buildings.
- Few RC buildings suffer damage grade 2.

X Very destructive

- General panic. Most people are washed away.
- Most large vessels are moved violently ashore, many are destroyed or collide with buildings. Small boulders from the sea bottom are moved inland. Cars overturned and drifted. Oil spill, fires start. Extensive ground subsidence.
- Damage of grade 4 in many masonry buildings.
- Few RC buildings suffer damage grade 3. Artificial embankments collapse, port water breaks damaged.

XI Devastating

- Lifelines interrupted. Extensive fires. Water backwash drifts cars and other objects in the sea. Big boulders from the sea bottom are moved inland.
- Damage of grade 5 in many masonry buildings.
- Few RC buildings suffer damage grade 4, many suffer damage grade 3.

XII Completely devastating

- Practically all masonry buildings demolished.
- Most RC buildings suffer damage of at least grade 3.

1.6 THE DECEMBER 26, 2004 DEVASTATING TSUNAMI IN THE INDIAN OCEAN

The great Sumatra earthquake of December 26, 2004, measuring moment-magnitude around M_w 9.2 (see Box 1.2), caused one of the largest and more disastrous tsunamis ever experienced. An estimated 230,000 people lost their lives. The tsunami directly affected 16 nations all around the Indian Ocean and indirectly many other nations around the globe given that some thousands of European, Americans, Asians and other tourists/visitors lost their lives. The highest death toll, of about 130,000 was reported from Banda Aceh and Meulaboh along the northwestern coast of Sumatra, where tsunami run-up heights exceeded 30 m (Tsuji et al, 2005; Borrero et al., 2006). Within hours the tsunami propagated to all parts of the Indian Ocean, affecting Thailand, Sri Lanka, India, Maldives, and as far as east Africa (Figure 1.5). Extensive damage was caused to buildings of several types, such as houses, schools, hospitals, and government buildings, as well as to infrastructure, such as harbors, bridges, and railway (Figures 1.6–1.8).

BOX 1.2 The great tsunamigenic earthquake of December 26, 2004 in Sumatra

The great earthquake of December 26, 2004, 7:58 A.M. local time, was among the largest earthquakes measured in the Earth. It ruptured about 250 km to the west of Sumatra at a length of about 1200–1300 km along the Sumatra–Nicobar–Andaman megathrust system with an average source duration of around 500 s and a mean velocity of 2.4 km/s (Lay et al., 2005; Kanamori, 2006; Lambotte et al., 2006; Vallée, 2007). The moment magnitude, M_w, estimates range from 9.0 to 9.3, the upper value, corresponding to a moment of 1.3×10^{30} dyne-cm, being quite possible according to the analysis of Earth's low-order normal modes by Stein and Okal (2005). Nearly 250,000 km^2 of oceanic rock was ruptured, a large portion of it slipping 10–15 m or more along the surface of the fault. The tsunami magnitude, M_t = 9.1 suggests that the overall size of the tsunami is consistent with what is expected of an earthquake of that size.

On March 28, 2005, an M_w = 8.6 tsunamigenic earthquake ruptured near Nias Island on the southeastern extension of the December 26, 2004 rupture zone. The magnitude, M_t = 8.5 (Kanamori, 2006), of the rather small tsunami was again consistent with the size of the causative earthquake.

About 3 months later, a second very large earthquake of M_w = 8.6 ruptured to the south extension of the rupture zone of December 26, 2004 (see Box 1.1). The tsunami caused was smaller than the one of December 26, 2004 but it was powerful enough to kill approximately 900 people and to render 22,000 people homeless (Figures 1.9–1.11).

The December 26, 2004 great tsunami drastically affected the development of tsunami studies and risk mitigation actions on a global scale.

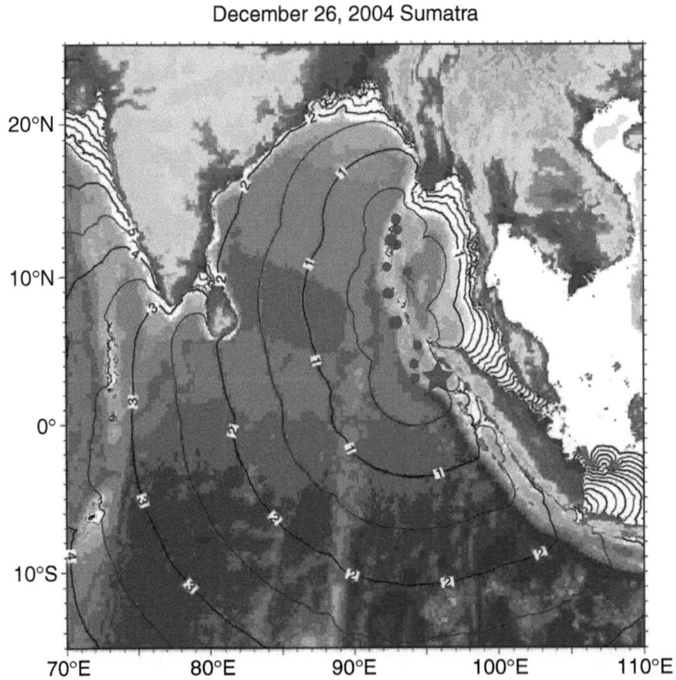

Figure 1.5 ***Travel times (in hours) of the Sumatra 2004 big tsunami.*** Star and solid circles illustrate epicenters of the main shock and major aftershocks, respectively. *(Calculated by K. Satake, Earthquake Research Center, Tokyo University.)*

Figure 1.6 ***Destruction in the coastal zone of Padong, Phuket Island, Thailand, after the strong inundation of the Sumatra 2004 tsunami.*** Photo taken 1 h after the inundation. *(Photo courtesy by A. Loukatzikou.)*

Figure 1.7 ***Destruction in the coastal built environment of Galle city, Southern Sri Lanka, where the height of the Sumatra December 26, 2004 tsunami was measured at 10 m (Papadopoulos et al., 2006).*** Picture was taken on January 10, 2005. *(Photo courtesy by S. Pavlides, University of Thessaloniki, Greece.)*

As soon as the tsunami struck the international tsunami community mobilized coordinated post-event field surveys in nearly all the nations that were struck by the tsunami in the Indian Ocean. Until that event, the Pacific Tsunami Warning System was the only regional system operating in the globe, thanks to the close collaboration of many nations around the Pacific Ocean under the coordinating umbrella of IOC/UNESCO since 1968. In 2005, the assemblies of country members of IOC/UNESCO decided

Figure 1.8 ***Fishing boats moved ashore in the coast of Galle city, southern Sri Lanka, due to the Sumatra 2004 tsunami.*** Picture was taken on January 10, 2005. *(Photo courtesy by G.A. Papadopoulos.)*

to establish regional tsunami warning systems in the Indian Ocean, in the North East Atlantic, the Mediterranean Sea, and its connected seas as well as in the Caribbean Sea (Figures 1.12 and 1.13).

Coordinated research projects were also initiated, for example, the European Commission funded several research projects including the pan-European tsunami project TRANSFER, and the SAFER project dedicated to seismic early warning. In parallel, dedicated working groups

Figure 1.9 ***Destruction of the front wall of a brick house in Sri Lanka.*** Picture was taken on January 9, 2005. *(Photo courtesy by G.A. Papadopoulos.)*

Figure 1.10 ***Destruction of a two-storey brick house in Sri Lanka.*** Picture was taken on January 9, 2005. *(Photo courtesy by G.A. Papadopoulos.)*

Figure 1.11 ***Failure of railway due to lateral spreading and collapse of its embankment caused by the soil erosional action of the Sumatra 2004 tsunami.*** Picture taken in Abalangoda, south west Sri Lanka, on January 10, 2005 (after Papadopoulos et al., 2006). The tsunami wave moved from the right-hand side to the left-hand side of the picture. *(Photo courtesy by G.A. Papadopoulos.)*

of the Tsunami Commission of IUGG organized valuable databases of instrumental records as well as collections of post-event field observations regarding the December 26, 2004 tsunami of the Indian Ocean.

The list of papers, reports, books, and conference proceedings published for the December 26, 2004 and March 28, 2005 earthquakes and tsunamis

Figure 1.12 ***Indoor water level mark after the 2004 Sumatra tsunami.*** Picture taken in Hambantota, southeast Sri Lanka, on January 12, 2005. Using this water level mark Papadopoulos et al. (2006) measured maximum wave height of ca. 11 m above m.s.l. *(Photo courtesy by G.A. Papadopoulos.)*

is very long. Some papers and reports can be found in several collecting volumes, including the ones edited by Papadopoulos and Satake (2005), Tinti (2005), Iwan (2006), Murty et al. (2007), Satake et al. (2007) and Bernard and Robinson (2009) (Figures 1.14 and 1.15).

1.7 THE 2011 DEVASTATING TSUNAMI IN JAPAN

On March 11, 2011, at 14:46 local time or universal time +9, a very large earthquake measuring magnitude M_w 9.0 ruptured off the North East shore of Japan in the Pacific Ocean (see Box 1.3). The earthquake generated an

Figure 1.13 ***The author (right) talking with the manager of a coastal hotel near Kalutara (South West Sri Lanka) about the water level mark left on trees from boats moved ashore by the Sumatra 2004 tsunami.*** Picture taken on January 9, 2005. In Kalutara maximum wave height of ca. 11 m above m.s.l. was measured by Papadopoulos et al. (2006). *(Photo courtesy by S. Pavlides.)*

equally large tsunami which devastated the North East coastal zone of Japan, particularly in Tohoku region where the wave height reached up to about 40 m (Mori et al., 2011), while the tsunami penetrated inland up to about 5 km. An estimated 19,508 people lost their lives including missing persons, nearly 90% of them due to the tsunami (Ando et al., 2011). Within hours the tsunami propagated to all directions of the Pacific Ocean affecting places as far away as California, where damage was noted in Crescent City (Figures 1.16 and 1.17)

BOX 1.3 The great tsunamigenic earthquake of March 11, 2011 in Japan

The great (M_w 9.0) megathrust earthquake of March 11, 2011, 14:46 local time, known as the Tohoku-Oki earthquake, was among the largest earthquake events ever measured in the Earth. It ruptured offshore of east Japan about 100 km to the east of North East Honshu Island at a length of about 500–550 km (Ammon et al., 2011; Suzuki et al., 2011) along the Pacific and Eurasian interplate boundary on a low-angle seismic plane dipping 10° to 12° (Duputel et al., 2011; Nettles et al., 2011). The average fault displacements were of ~15–20 m (Lay et al., 2011), while the region with large slip was approximately 150-km wide by 300-km long, with source duration of 150–160 s, a mean velocity of 1.5–2.5 km/s, maximum slip estimates ranging from 35 m to 60 m and seismic moment of 3.8–4.42 × 10^{22} Nm (Ammon et al., 2011; Fujii et al., 2011; Iinuma et al., 2011; Suzuki et al., 2011). However, the region with large slip was relatively compact compared with the 200-km wide by 500-km long aftershock area (Ammon et al., 2011) but the size of the main fault, being about 450 km in length and 200 km in width, was consistent with the size of the aftershock area (Yoshida et al., 2011). Compared to the December 26, 2004 Sumatra earthquake, the 2011 Japan earthquake was characterized by shorter duration of HF energy radiation and larger displacement amplitude (Hara, 2011).

Nearly complete stress drop was suggested by the stress-field analysis based on focal mechanisms of earthquakes and GPS records in the vicinity of the 2011 Tohoku-Oki big earthquake (Hasegawa et al., 2011). Such a result is consistent with the supercycle earthquake model proposed for the occurrence of earthquakes off Tohoku and elsewhere in other megathrust structures in subduction zones, such as the Cascadia and the Sumatra ones (Sieh et al., 2008; Kulkarni, 2013; Goldfinger et al., 2013).

Before the strong ground motion hit cities, the JMA issued Earthquake Early Warning (EEW) announcements to the general public of the Tohoku district and then the warning was automatically broadcast through TV, radios and cellular phone mails (Hoshiba et al., 2011). However, for the Tokyo region, JMA EEW expected intensities of 4, which was an underestimation of the observed intensity (5-upper), likely due to the large extent of the fault rupture.

Destruction of infrastructure and economic impact such as that caused in North East Japan by the 2011 tsunami had never been reported in the past. For example, in Fukushima, nuclear power plants were drastically affected by the tsunami resulting in the meltdown of three reactors. However, the Onagawa nuclear power plant, located 15 m above sea level, underwent a small amount of tsunami inundation, but there was no damage to the reactor buildings and equipment, which shut down safely following the earthquake (Somerville, 2012). In Sendai, massive tsunami inundation occurred in the coastal airport which remained closed for about 1 month. Problems related to the tsunami early warning procedures and the response of the public to evacuation alerts are discussed later.

Figure 1.14 ***Sand layer left behind at distance of about 300 m inland near Batticaloa, western Sri Lanka, by the Sumatra 2004 tsunami.*** Picture taken on January 13, 2005. Maximum wave height of ca. 7 m above m.s.l. was measured by Papadopoulos et al. (2006). *(Photo courtesy by S. Pavlides.)*

The 2011 tsunami was the first to provide so many instrumental records. In addition to the tide gauge and global positioning system (GPS) records, ionospheric disturbances were recorded by ionosondes in Japan and Taiwan (Liu and Sun, 2011) as well as by GPS-derived Total Electron Content techniques applied to the Japanese GPS network GEONET (Rolland et al., 2011). On the other hand, the passage of the tsunami was recorded by cabled ocean bottom tsunami sensors offshore Japan (Maeda et al., 2011), by Deep-ocean Assessment and Reporting of Tsunami (DART®) sensors in the Pacific Ocean (Lay et al., 2011) as well as by a network of deep-seafloor electromagnetometer network near Japan (Sugioka et al., 2014). The tsunami was also recorded by a high frequency (HF) ocean surface radar installed in the eastern coast of the

Figure 1.15 ***Fish left behind in the coastal zone of Padong, Phuket Island, Thailand, after the inundation of the Sumatra 2004 Tsunami.*** Photo taken 1 h after the inundation. *(Photo courtesy by A. Loukatzikou.)*

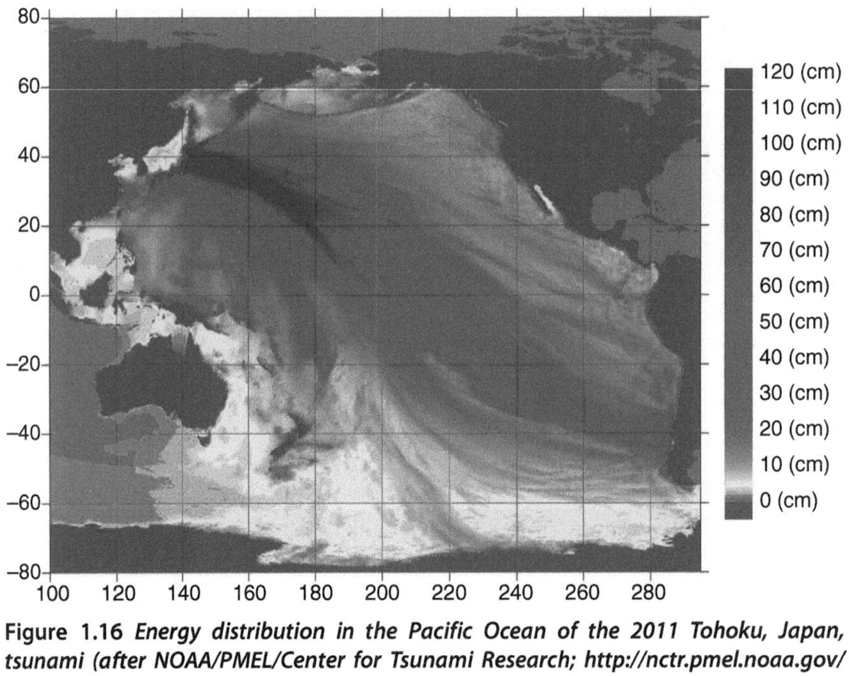

Figure 1.16 ***Energy distribution in the Pacific Ocean of the 2011 Tohoku, Japan, tsunami (after NOAA/PMEL/Center for Tsunami Research; http://nctr.pmel.noaa.gov/ Honshu 20110311/.)***

Kii Channel, at a range of about 1000 km from the epicenter along the eastern to southern coasts of Honshu Island (Hinata et al., 2011). It is of interest that the WERA Ocean Radar in Chile observed tsunami signatures after the earthquake in Japan on March 11, 2011 (HELZEL GmbH Press Release, 2011).

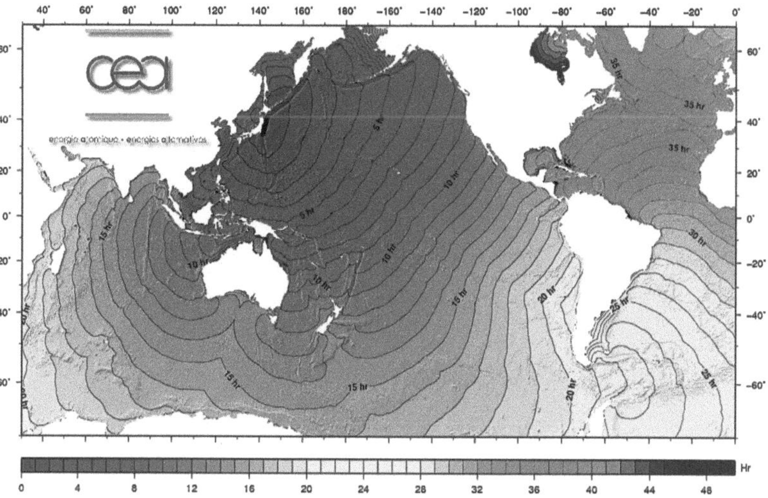

Figure 1.17 ***Travel time modeling for the 2011 Tohoku Tsunami (CEA, France, http://www.emsc-csem.org/Earthquake/196/Mw-9-0-off-the-Pacific-coast-of-Tohoku-JapanA-Earthquake-A-on-March-11th-2011-at-05-46-UTC).*** Time contours in hours.

Deep ocean records help us to better understand tsunami propagation and how ocean bathymetry may affect the configuration of tsunami hazard along coastal zones. But the utilization of deep ocean real-time records is of crucial importance for operationally effective early warning purposes. Observations from HF ocean surface radars add a new role to such devices for measuring the detailed surface current fields with high spatiotemporal resolution toward understanding detailed processes of resonant response to tsunami waves in coastal regions. On the other hand, ionospheric and electromagnetic records constitute new promising tools for improving real-time tsunami detection and forecasting. Ionospheric anomalies were observed in relation to the megaearthquake of 2011 as well (Liu, et al., 2011; Maruyama et al., 2011).

For such a large tsunami to occur it is generally believed that its generation should be attributed to the large fault displacement at the source. However, Kawamura et al. (2012) assuming a conservative displacement of 20 m and adopting that the thrust fault rapidly deformed the seafloor (Ide et al., 2011), supported that the tsunamis should be generated all along the axis of the Japan Trench as shown by the scenario of Tanioka and Satake (1996). Then, based on high-resolution topographic surveys and detailed seafloor observations, those authors suggested a new scenario for additional tsunami generation from large submarine landslides in the Japan Trench

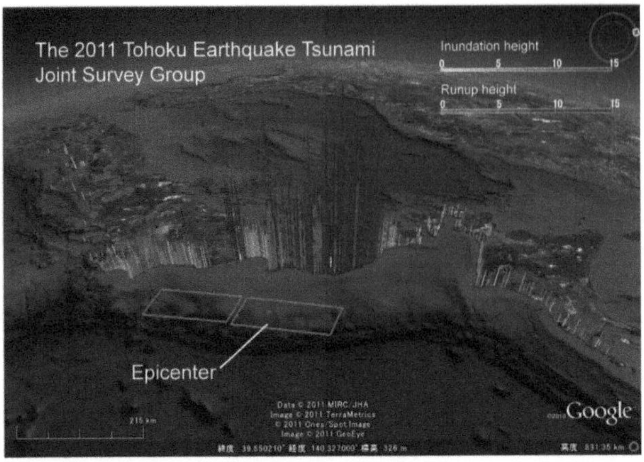

Figure 1.18 ***Image of the Japan convergent margin, with run-ups from the March 2011 tsunami.*** White line, convergent margin; vertical lines, run-up heights. *(Image credits: Google Earth background and Tohoku Joint Survey Group.)*

set in motion by the earthquake. According to Kawamura et al. (2012) this hypothesis should explain the relation between large displacement of the thrust fault and generation by the 2011 and 1896 Tohoku earthquakes (Figure 1.18).

The study of tsunami sediment deposits in the plain of Sendai along with historical evidence indicated that in AD 869 the area was inundated by the so-called Jōgan very strong tsunami (Minoura et al., 2001) (Figure 1.19). The tsunami penetrated inland about 4 km. From seismotectonic considerations and tsunami numerical modeling, Minoura et al. (2001) concluded that the tsunami was triggered by a large-scale earthquake having its source exactly at the area where the 2011 earthquake ruptured. In their prophetic article, those authors estimated that the recurrence of the Jōgan event is about 1000 years and that the AD 869 tsunami was larger than tsunamis generated by normal earthquakes there and concluded that (a) the possibility of a large tsunami striking the Sendai plain is high and (b) a tsunami similar to the Jōgan one would inundate the present coastal plain for about 2.5–3 km inland.

After the 2011 event, the study of ostracodes assemblages collected from the tsunami sediment deposits left behind 1 km inland in Rikuzentakata City showed that the sediment was derived from the seafloor from at least 9 m water depth (Tanaka et al., 2012). The authors of that paper suggested that such studies may provide insight into past tsunami wave height and potentially earthquake slip and magnitude.

「陸奥国地大震動。流光
如晝隠映。頃之。人民叫
呼。伏不能起。或屋仆壓
死。或地裂埋殪。馬牛駭
奔。或相昇踏。城郭倉庫
。門櫓墻壁。頽落顛覆。
不知其数。海嘯哮吼。聲
似雷霆。驚濤涌潮。泝洄
漲長。忽至城下。去海数
千百里。浩々不辨其涯涘
矣。原野道路。惣為滄溟
。乗船不遑。登山難及。
溺死者千許。資産苗稼。
殆無孑遺焉。」

Figure 1.19 ***Historical Japanese document describing the large AD 869 Jōgan earthquake and tsunami.*** *(Provided by K. Minoura, Tohoku University, Japan.)*

The economic impact of the Tohoku tsunami and earthquake was global. In the aftermath of the catastrophic 2011 event all nuclear power plants were shut down within Japan. This resulted in power shortages of about 30% of the total national need which in turn led to the need for importing fossil fuels which had negative consequences for the Japanese economy (Somerville, 2012). On the other hand, outside Japan the Fukushima disaster marked the beginning of huge reduction in nuclear power generation around the world, for example in Germany, Switzerland, and Italy. The impact of this global transition from nuclear to fossil fuel power generation may mark a serious setback in the struggle to control global warming (Somerville, 2012). (Figures 1.20 and 1.21).

Figure 1.20 ***Tohoku tsunami 2011.*** **Boat moved atop a building.** *(Photo courtesy by G.A. Papadopoulos.)*

Figure 1.21 ***Tohoku tsunami 2011.*** **Metallic building after the tsunami attack.** *(Photo courtesy by G.A. Papadopoulos.)*

1.8 TSUNAMI EARLY WARNING SYSTEMS

1.8.1 Requirements, Technologies, Current Practices

The challenges for disseminating an efficient warning well before the arrival of the first tsunami wave have been the subject of several discussions among the scientific, technological, and civil protection communities, particularly after the experience of the big Sumatra 2004 and Tohoku (Japan) 2011

Table 1.4 Typical response times as a function of the local, regional or distant nature of the tsunami (http://itic.ioc-unesco.org/images/docs/TWCoverview_apr08.pdf)

Tsunami type	Typical time to impact	TWC response time
Local	0–1 h	2–5 min
Regional	1–2 h	5–10 min
Distant	>2 h	10–20 min

tsunami events. Usually the terms Tsunami Warning Systems and Tsunami Early Warning Systems are in use depending on the time available to a Tsunami Warning Center (TWC) to warn people. A typical scheme of response times is shown in Table 1.4. To accomplish tsunami detection, a TWS should integrate sensor networks to determine in very short time the parameters (time, location, magnitude) of a strong earthquake susceptible to produce tsunami. The routine procedure to do so is based on seismograph records. As soon as such an earthquake has been detected and analyzed either automatically or/and manually, a tsunami forecasting estimation is produced based on databases of presimulated tsunami events. That is, as soon as the earthquake is determined from its parameters as being potentially tsunamigenic, the tsunami scenario which is closest to the earthquake parameters is automatically retrieved from the database. The information contained in the tsunami message includes the earthquake parameters (origin time, epicenter, magnitude) and a set of forecasting elements such as the estimated tsunami arrival times and possibly the estimated tsunami wave heights in certain forecast points in coastal sites of interest. The tsunami bulletin (message) is issued and disseminated to prescribed recipients, such as civil protection and other authorities, other TWCs etc. Means of telecommunication usually include email, fax, telephone, and the Global Telecommunications System (GTS) which is in routine use by the meteorological agencies worldwide.

If it is not possible to use the database of presimulated tsunami scenarios, regardless the reason, or such a database is not available, then a tool comprising a set of empirical rules called the *Decision Matrix* is an alternative. The Decision Matrix provides three or four levels of tsunami alert depending on the earthquake parameters. The higher the magnitude the higher the level for tsunami alert. However, no forecasting details are produced, but only a scaling incorporating 3 or 4 alert levels is given, for example, *Tsunami Information*, *Tsunami Advisory*, or *Tsunami Watch*. Such a scaling depends not only on the characteristics of the earthquake, e.g., magnitude, but also on the distance of a particular coastal site from the earthquake epicenter. Therefore, for a given earthquake event the tsunami alert could be at the level of Tsunami

Watch for a set of forecast points and at the level of Tsunami Advisory or even only at the level of Tsunami Information in other localities. By applying tools for the tsunami travel time calculation the arrival times of the wave in preselected forecast points could be included in the tsunami messages. In Chapter 7, the reader can find the Decision Matrices which are currently in use by TWCs in the Mediterranean and the North East Atlantic regions.

Verification that a tsunami has been generated comes from instrumental records and or eyewitnesses in the coastal zone. Classic instruments recording tsunamis are tide gauges in shorelines that transmit signals of the sea level changes via Internet, satellite, mobile telephony, or other technologies. To ensure early detection of tsunamis and to acquire data critical to real-time forecasts, the National Oceanic and Atmospheric Administration (NOAA), USA, has placed DART stations at sites in regions with a history of generating destructive tsunamis. NOAA completed the original 6-buoy operational array in 2001 and expanded to a full network of many stations in March 2008 installed in the Pacific and Indian oceans as well as in the Caribbean Sea (http://www.ndbc.noaa.gov/dart.shtml). In Japan, a few sea-bottom tsunameters existed when the big tsunami of 2011 occurred. After this catastrophic event the NIED (National Research Institute for Earth Science and Disaster Prevention; http://www.bosai.go.jp/e/) took the lead and developed a network of 150 seafloor observatories along the Japanese Trench each one comprising seismometers and hydro-pressure gauges. The seafloor observatories are connected by fiber optic cables of 5,800 km in total length, thus making early warnings of seismic and tsunami events possible. A similar, large-scale network called DONET was organized by JAMSTEC. The main purpose of DONET is to monitor the hypocentral region of the big Tonankai earthquake that is predicted to occur with a probability of more than 70% within the next 30 years according to the report published by the Earthquake Research Committee. DONET consists of an approximately 300 km length of backbone cable system, 5 science nodes, and 20 observatories. The installation of 20 stations at Kumanonada started in 2006 and was completed in August 2011. The seismic data are provided to the Japan Meteorological Agency (JMA) and the NIED, where the data will be used for the earthquake early warning.

Current operational weaknesses of tsunami warning centers include an inability to detect landslide and volcanic sources as possible agents of tsunami generation, a high rate of false alarms reducing the confidence of citizens to warning systems and an inability to provide early enough warnings for local tsunamis except in a few areas.

1.8.2 Near-Field Tsunami Early Warning

In the near-field (or local) domain, however, the time constraints for early warning are extreme. From the tsunami risk point of view it is of great importance, as according to global tsunami statistics, about 80% of victims due to tsunamis are caused within the first 1 h of tsunami propagation (Gusiakov, 2009). The need for tsunami early warning in the near field has been discussed internationally (Schindelé, 1998; Murata et al., 2010). One definition for near-field tsunamis is that the wave travel distances are of a similar order (of magnitude) to the earthquake rupture length (Lauterjung et al., 2010). Such physical preassumptions were adopted for the development of the German Indonesian Tsunami Early Warning System (GITEWS) in the aftermath of the Sumatra 2004 event on the basis of the correct characterization of the earthquake rupture, including the seismic slip distribution, based on seismological and GPS records. Information of this type is available 5–10 min after the event at best.

In near-field conditions, the operative efficiency of a tsunami early warning system depends on the times needed for seismic signal communication, t_S, and for evacuation, t_E (Sasorova et al., 2008). Therefore, the next relation should be realized:

$$t_{tr} > t_S + t_E \tag{1.9}$$

where t_{tr} is the first tsunami wave travel time from the source to the closest settlement. However, each one of the parameters t_S and t_E in formula (1.9) is composed of more than one time components. Seismic signal communication incorporates the time needed for the determination of the earthquake focal parameters, t_{SD}, the time needed for the tsunami decision-making, t_{TD}, the time needed for the transmission of the earthquake information, t_{ST}, as well as the result of tsunami decision to the operational center of civil protection. In addition, the time for evacuation incorporates another two components. The first is the time, t_{TW}, needed to transmit warning information from the civil protection to the population. Finally, after transmitting this information there is need to allow for some time to respond for real evacuation, t_{EV}. Assuming that under optimum conditions the earthquake information and the tsunami decision are transmitted to the civil protection authority automatically as soon as the earthquake determination has been performed, then we get:

$$t_{tr} > t_{SD} + t_{TW} + t_{EV} \tag{1.10}$$

We assume that the time, t_{SD}, needed for the earthquake determination is 3–5 min at minimum, and the time, t_{TW}, needed for the civil protection authority to issue and disseminate an early tsunami warning is another 5 min at best. The time t_{EV} for the population to receive the warning information and understand it correctly and to run away for real evacuation takes 15 min at best. Summing these we get $t_{tr} > 23$ min. However, in near-field conditions, the time component t_{tr} is on the order of 15 min. For example, this was exactly the 2004 case of Banda Aceh in Sumatra. Also, this is the typical case for many coastal sites, which are threatened by near-field tsunami sources in the North East Atlantic and the Mediterranean Sea region. This result underlines the urgent need to drastically compress the time needed for warning and real evacuation by developing local tsunami early warning systems. Such a system was developed in Rhodes Island, Aegean Sea, Greece, within the frame of the project NEARTOWARN supported by DG ECHO of EU. This pilot system is further described in Chapter 7.

1.9 LESSONS LEARNED FROM NEAR-FIELD EARLY WARNINGS: THE JAPANESE EXPERIENCE

1.9.1 Tsunami Arrival in 25 Min

To evaluate how a tsunami early warning system operating in near-field conditions can satisfy formula (1.10), the great Tohoku tsunami that hit east Japan on March 11, 2011 (14:46, Japan time) was taken as a reference case. According to Kamigaichi (2012), a seismic early warning for an earthquake of magnitude $M = 7.9$ was issued by the JMA through the public broadcasting system of Nippon Hoso Kyokai (NHK) about 1.5 min after the earthquake generation. A decision that a large tsunami of 3–6 m height was likely was derived from a database of presimulated tsunami scenarios and early warning was disseminated through the NHK system 3 min after the earthquake generation. A first sea level rise of about 0.2 m arrived in some localities within 10–15 min but its nature was not clear. Very likely this first arrival did not represented the big tsunami that arrived later but was the result of massive coseismic subsidence of the east coast of North East Honshu (Figure 1.22).

The first tsunami wave arrived about 25 min after the earthquake origin time and in some coastal localities it was of much higher amplitudes than the forecasted ones. This means that the first warning was issued on time but underestimated dramatically the tsunami height, which is explained by the fact that the first earthquake magnitude was also underestimated due

Figure 1.22 ***Tohoku tsunami 2011.*** Typical tsunami evacuation area at higher floors. *(Photo courtesy by G.A. Papadopoulos.)*

to saturation of the seismic records. As soon as real tsunami records from tsunameters as well as broadband seismic records became available, JMA revised the tsunami warning about 28 min after the earthquake origin time. The revised tsunami warning parameters fitted the wave heights and arrivals observed much better. In fact, JMA warned for 10-m wave for Miyagi and 6-m wave for Iwate and Fukushima.

As regards the time of the first tsunami warning by JMA, one may argue that the warning was successful. In fact, by ignoring the last term in formula (1.10), which refers to the time needed for population evacuation, we get 25 min $\gg$ 3 min. However, the evacuation of population only partly performed well. Several populations never evacuated, while others were delayed in evacuation even in coastal segments where the first tsunami arrived about 1 h after the earthquake origin time. Problems with

Figure 1.23 ***Tohoku Tsunami 2011.*** **Big boat moved ashore 1 km inland.** *(Photo courtesy by G.A. Papadopoulos.)*

the evacuation were often due either to the content of the warning message itself or to the announcement that the first wave observed was low. For example, Kamigaichi (2012) reported that the initial warning for an expected tsunami of 3 m caused delay in evacuation since many residents considered that they were safe. On the other hand, the announcement that the first tsunami height observed was only 0.2 m caused also delays or even interruptions in the evacuation procedures. The negative response to evacuation resulted in a dramatic increase in the number of victims (Figure 1.23).

The case of Natori city, examined in detail by its mayor Sasaki (2012), is extremely illuminating as regards real problems regarding the transmission of tsunami warning information to local communities and effective evacuation. In Natori, the first tsunami arrival was noted on 15:51 with wave height exceeding 10 m. However, due to the earthquake shaking, the municipal disaster management radio communication network suffered from power supply short-circuit, while the local TV system suffered blackouts. The warning means that performed well included mobile radio, public information by firefighting team vehicles, public information provided by the neighborhood association, and the voluntary disaster prevention organization as well as calling from neighbors. In addition, the local sound machine did not sound since some of its metal pieces fell onto the power supply of the radio transmitter on the roof, due to the strong earthquake shaking, and then it short-circuited (Figure 1.24).

Figure 1.24 ***Tohoku tsunami 2011.*** Coastal buildings destroyed by the wave. *(Photo courtesy by G.A. Papadopoulos.)*

After these experiences, the mayor in Natori recommended that in order to get communication means, which will be usable at the time of a great disaster, one might rely on low technology rather than on high technology (Sasaki, 2012). This is consistent with one of the lessons learned according to Koshimura (2012) who concluded that there are still limitations on the reliability of science and technology, which is in use for the tsunami early warning in near-field conditions. He supported that in these conditions the tsunami warning information is useful to let residents know that they are in danger, but it does not guarantee their safety. He also emphasized that under such conditions there is no need to wait for official information, a practice that certainly could compromise the evacuation time. From this point of view, a perfect example was presented by Suenaga (2012) who showed that our own decisions and actions based on correct information and knowledge could save our lives through the dogma *tsunami ten-den-ko*, that is, *save your life by yourself*. He presented an excellent example from the Kamaichi city where many students evacuated very effectively based on their good education and training as well as on their own action and *tsunami ten-den-ko* practice at the time of the tsunami threat.

It should be noted, however, that the tsunami *ten-den-ko* practice does not reduce the value of the early warning systems and does not substitute such systems for two reasons. The first is that the *ten-den-ko* practice falls in the response stage following the warning. Of course people may feel the earth shaking and react by running to higher grounds before the issue of a warning signal. This practice certainly may save lives and is highly recommended. However, the earth shaking is not always strong enough to warn people, particularly in cases of tsunami earthquakes where the tsunamis are disproportionately large with respect to the earthquake size (see Section 1.3). Therefore, the tsunami *ten-den-ko* practice should be viewed only in synergy with the warning systems.

From the 2011 Japanese experience, briefly outlined earlier, one may conclude that in the chain *determine physical parameters–warn people–evacuate people*, there are at least two seriously weak links. The first concerns the large uncertainties involved in the earthquake parameter determination and in the parameters of the expected tsunami. The second is about the very practical implementation of the system that is about the response of population to evacuation. To improve the Japanese tsunami warning system for local tsunamis, JMA decided to revise and make simpler the procedure as explained officially and in details in the report of Kamigaichi (2012). According to this report, the first tsunami warning should be disseminated in 3 min. In case the magnitude of the earthquake is underestimated, the first warning should be based on the assumed maximum magnitude of the area, while the estimated tsunami amplitude should be mentioned only qualitatively as an emergency message. Considering the scatter of tsunami amplitude involved but also for closer linkage of warning to hazard maps, it was also decided to reduce the number of levels of estimated tsunami amplitude from 8 to 5. Finally, even though the first observed tsunami amplitude is too small it should not be reported in numbers aiming to avoid underestimation of the threat by the residents (Figure 1.25).

One of the critical lessons learned from the Tohoku 2011 experience is the underestimation of the earthquake size and because of this of the tsunami size too. Another critical point is that many residents did not evacuate because of misunderstanding of the warning messages. The third is that for those who decided to evacuate in response to the early warning evacuation was not an automatic procedure. On the contrary, the time needed for real evacuation was much longer than was expected. In view of these serious problems, the decision of JMA to revise warning procedures making them simpler and more qualitative than they were before looks quite realistic. On the other hand, looking with a critical eye at the population response to the tsunami warning in Japan on March 11, 2011, one should consider that thousands of lives were saved precisely because they evacuated following the warning signals.

Figure 1.25 ***Tohoku tsunami 2011.*** Memorial stone established after the tsunami disaster. *(Photo courtesy by G.A. Papadopoulos.)*

1.9.2 Tsunami Arrival in 5 Min

A good example of a very local tsunami comes from the Okushiri Island tsunami case of July 13, 1993 in the north side of the Japan Sea. An earthquake measuring $M = 7.2$ occurred in Japan Sea at a distance of only 70 km offshore west of Hokkaido Island at 22:17. At 22:22, the NHK system announced the earthquake focal parameters along with a tsunami early warning message. The first, destructive tsunami wave arrived just at the announcement time practically leaving time for evacuation not due to the warning but only due to the *tsunami ten-den-ko* practice. In fact, many residents reacted within only a few minutes as they felt the strong earth shaking. The new practice introduced by JMA after the Tohoku event of 2011 certainly improves the early tsunami warning capabilities for tsunamis arriving only within 5 min from the earthquake origin time.

CHAPTER 2

Historical and Geological Evidence of Tsunamis in Europe and the Mediterranean

2.1 INTRODUCTION

The very small number of tsunamis that are known from instrumental records so far creates serious difficulties in developing standard and low-uncertainty statistic and probabilistic approaches for the tsunami hazard estimation in a way similar to that applied in probabilistic seismic hazard assessment. Therefore, the identification of past tsunami events, either from historic documentary sources or by means of onshore and offshore geological methods (paleotsunamis), by sedimentary records, is of particular value to extend the tsunami time series into the historical and geological past. In a few cases, archeological observations also provide evidence for the identification of paleotsunamis. Figure 2.1 illustrates the main tsunamigenic zones in the Mediterranean and adjacent seas determined from historical, geological and instrumental data (Papadopoulos et al., 2014a). The zones are classified according to their tsunami potential.

In this chapter the historical, geological, archeological and instrumental records of tsunami events in the European–Mediterranean region are critically described. The assignment of intensities of the tsunami events is based on the 12-point Papadopoulos and Imamura (2001) tsunami intensity scale unless otherwise indicated.

2.2 HISTORICAL SOURCES

The occurrence of tsunamis in the European–Mediterranean region has mainly been known through a large number of documentary sources stretching over a long period of time, such as histories and chronicles, descriptions by travelers, reports of authorities, and eyewitness accounts. Such documentary sources have provided a good basis for the publication of a variety of tsunami data compilations, such as descriptive and/or parametric tsunami cataloges and books. In modern times, descriptive tsunami cataloges have been published by several authors including Galanopoulos (1960),

Tsunamis in the European-Mediterranean Region. http://dx.doi.org/10.1016/B978-0-12-420224-5.00002-8

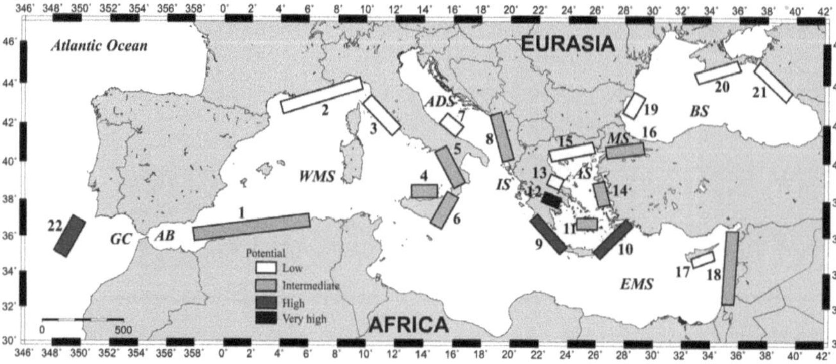

Figure 2.1 ***Tsunamigenic zones, defined from documentary sources, and their relative tsunami potential classification.*** WMS, western Mediterranean Sea; GC, Gulf of Cádiz; AB, Alboran Basin; EMS, Eastern Mediterranean Sea; AS, Aegean Sea; ADS, Adriatic Sea; MS, Marmara Sea; BS, Black Sea. Zonation key: 1, East Alboran Sea/North Algerian Margin Sea; 2, Liguria and Côte d'Azur; 3, Tuscany; 4, Aeolian islands; 5, Tyrrhenian/Calabria; 6, Eastern Sicily and Messina Straits; 7, Gargano Promontory; 8, East Adriatic Sea; 9, West Hellenic Arc; 10, East Hellenic Arc; 11, Cyclades; 12, Corinth Gulf; 13, Maliakos Bay; 14, East Aegean Sea; 15, North Aegean Sea; 16, Marmara Sea; 17, Cyprus; 18, Levantine Sea; 19, Bulgaria; 20, Crimea; 21, East Black Sea; 22, South West Iberia (Papadopoulos et al., 2014a).

Ambraseys (1962, 2009), Antonopoulos (1979, 1980), Tinti and Guidoboni (1988), Soloviev (1990), Guidoboni et al. (1994), Guidoboni and Comastri (2005), and Salamon et al. (2007, 2009). Parametric cataloges, containing parameters of both the tsunamis and their causative events (e.g., earthquakes) have been organized and published, by, among others, Papadopoulos and Chalkis (1984), Papazachos et al. (1985), Tinti and Maramai (1996), Altinok and Ersoy (2000), Soloviev et al. (2000), Papadopoulos (2003a, 2009, 2011), Yalciner et al. (2002), Tinti et al. (2004), Papadopoulos and Fokaefs (2005), Fokaefs and Papadopoulos (2007), Papadopoulos et al. (2007a, 2010), Baptista and Miranda (2009), Altinok et al. (2011), Maramai et al. (2014), Diakogianni et al. (2015).

The historical tsunami record in the Mediterranean region goes back to the Greek antiquity. Documentation of one of the earliest events was based on the historian Democles (fourth century BC). In his earthquake cataloge, he included a sea wave that attacked the coastal zone of Troy, modern North West Turkey, in about 1300 BC (Ambraseys, 1962; Antonopoulos, 1979; Papadopoulos and Chalkis, 1984; Soloviev et al., 2000; Papadopoulos, 2001). However, a relevant passage in Homer's *Iliad* referring to the *Wall War* has not attracted much attention so far. From Rapsodies M18–19 and M26–33, it becomes clear that a strong sea wave destroyed the wall that the Greeks constructed in the Aegean Sea coast to protect their fleet from the attacks of Troyans. Trees and wall materials were drifted away by the water; the ground was leveled by the strong wave; the coastal zone was covered by a sandy mantle while the sea caused the flow of rivers to invert. Although a conclusive interpretation about the nature of the strong sea disturbance seems unlikely, the Homeric description provides clues for interpreting the strong sea wave as a tsunami. In fact, effects such as the ones caused by the sea wave on the coast of Troy today are well recognized as being the result of strong tsunamis such as the ones of the Indian Ocean on December 26, 2014 and the Tohoku, Japan, tsunami of March 11, 2013. Therefore, a paleotsunami survey on the coast of Troy remains of great geo-scientific and archeological interest.

2.2.1 Eastern Mediterranean Basin

2.2.1.1 Hellenic Arc

The Hellenic Arc is a major geotectonic structure dominating the east Mediterranean basin and producing large earthquakes and tsunamis. Geological and archeological evidence collected in the Hellenistic/Roman harbor of Phalasarna, North West Crete (Pirazzoli et al., 1992; Dominey-Howes et al., 1998), as well as several documentary sources (Guidoboni et al., 1994; Ambraseys, 2009) imply that around AD 66 a tsunami flooded the western-southwestern part of Crete after a strong earthquake (see extensive critical review in Papadopoulos, 2011 and references therein).

The large tsunamigenic earthquake of AD July 21, 365 that likely ruptured the western segment of the Hellenic subduction zone (Figure 2.2) on a North East-dipping fault within the overriding plate (Shaw et al., 2008), is one of the most contentious and debated natural events in Mediterranean Sea history. However, the accounts of Marcellinus, Athanasius, and Jerome, which are the closest in time to the event, leave no doubt that a large area was affected since the tsunami propagated to the northwest, west, and south of the Hellenic trench and reached as far as Methoni in the South West Peloponnese, east Sicily, Alexandria and Dalmatia (see reviews in Jacques and Bousquet, 1984; Evagelatou-Notara, 1987/88; Guidoboni et al., 1994; Shaw et al., 2008; Papadopoulos, 2011). The AD 365 tsunami was one of the most important key events in the Mediterranean region. Notable tsunamis caused by strong earthquakes in the western Hellenic Arc were also observed on AD March 9, 1630; AD February 6, 1866; and AD September 20, 1867 (see review and references in Papadopoulos, 2011).

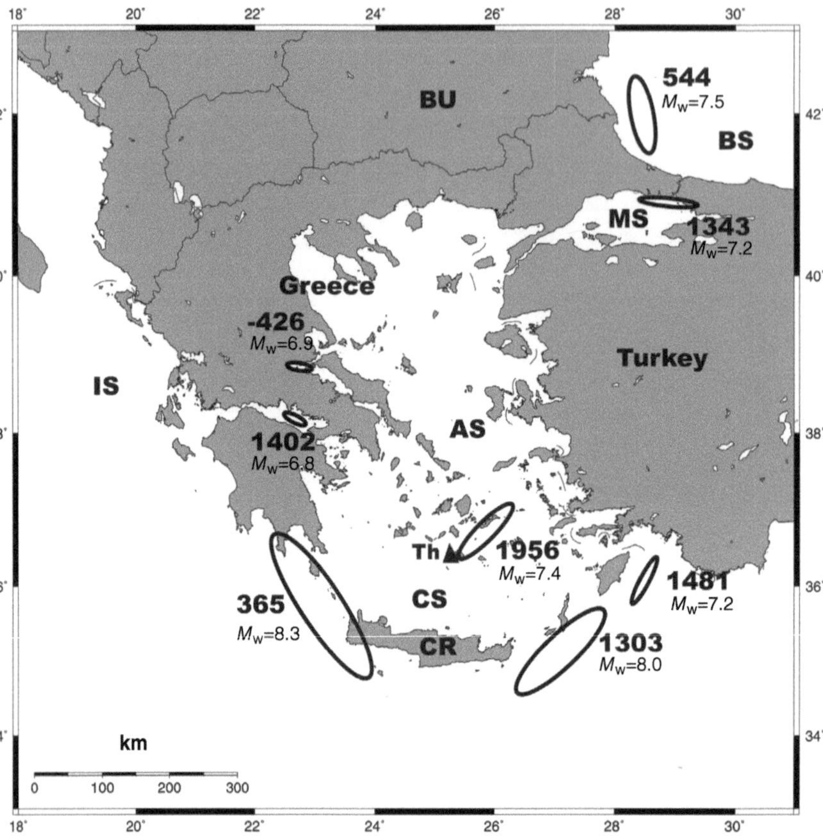

Figure 2.2 ***Source areas of the largest tsunamigenic earthquakes historically known in Greece, Turkey, and the surrounding regions.*** Key for geography: AS, Aegean Sea; BS, Black Sea; BU, Bulgaria; CR, Crete; CS, Cretan Sea; IS, Ionian Sea; MS, Marmara Sea. Symbol key: Figure near source area, year of earthquake occurrence; - , BC date; M_w, earthquake moment-magnitude. *(Slightly modified from Papadopoulos and Papageorgiou, 2014.)*

In the eastern segment of the Hellenic Arc, a long number of Byzantine, Venetian, and Arabic documentary sources indicate that the AD August 8, 1303 tsunamigenic earthquake, which very likely ruptured the arc between Crete and Rhodes (Figure 2.2), was one of the largest historically reported in the Mediterranean (see reviews in Guidoboni and Comastri, 1997; Papadopoulos, 2011). Similarly, the AD 1303 large tsunami also propagated in an extensive part of the eastern Mediterranean basin. The most serious destruction was reported from northeastern Crete where the large tsunami struck the capital city of Heraklion. The sea swept violently into the city with such force that it destroyed buildings and killed inhabitants. In Acre,

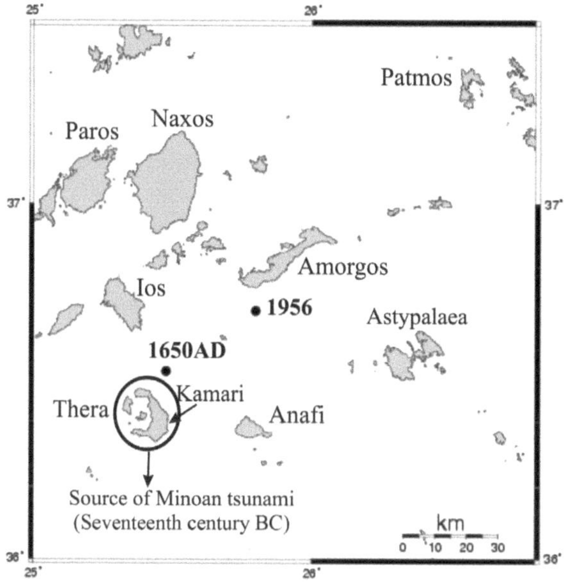

Figure 2.3 ***Important tsunamis in the Cyclades island complex, south Aegean Sea.***

Israel, people were swept away and drowned by the huge wave. In Alexandria, the wave destroyed port facilities. Both the AD 365 and AD 1303 tsunami key events are further discussed in next sections.

In the easternmost Hellenic Arc, strong tsunamigenic earthquakes occurred in AD 142 or 144, May 3, 1481 (Figure 2.2), April 1609 and January 31, 1741 in Rhodes Island, and in AD 556 in Cos Island. A local but still damaging tsunami occurred in Karpathos on AD February 9, 1948 (Papadopoulos et al., 2007a; Ebeling et al., 2012). As discussed later, trenches in late Holocene sediments at Dalaman, South West Turkey, revealed three tsunami sand layers attributed to the AD 1303, 1481, and 1741 tsunamis (Papadopoulos et al., 2005b, 2012a), which makes the Dalaman test site one of the key cases to test correlations between onshore tsunami sediment deposition and historical record of tsunamigenic earthquakes.

2.2.1.2 Aegean Sea

In the island complex of Cyclades, South Aegean Sea, the Late Bronze Age (LBA) or Minoan eruption of Thera (Santorini, Figure 2.3) was one of the most significant ever seen by humankind because of its very large size (VEI = 7+), its possible impact on LBA East Mediterranean civilizations, and for distributing huge amounts of tephra, thus creating an important marker horizon. According to relatively recent olive tree-ring radiocarbon dating the LBA event occurred in 1613 ± 13 BC (Friedrich et al., 2006). However, this dating has been considered as unreliable for a number of reasons (see Cherubini et al., 2014 and consequent debate in the same issue). The eruption history may have included four main phases (Heiken and McCoy, 1984) and concluded with the formation of the largest part of the caldera dominating the landscape of modern Santorini. The most intensive eruption phase lasted for about 3–4 days (Sigurdsson et al., 1990). The eruption of Thera is believed to have triggered a tsunami similar in magnitude or larger than the tsunami, which was generated by the AD 1883 Krakatau eruption in Sunda straits and rolled against the shores of Java and Sumatra with heights up to 35 m killing more than 36,400 people (Kawamata et al., 1993; Choi et al., 2003). A variety of geological and archeological field observations as well as results of numerical simulations are discussed in the next section in an effort to examine the generation mechanism of the so-called Minoan tsunami.

Another large tsunami was generated during the eruption of Columbo, a submarine volcanic edifice lying 7.3 km offshore to the northeast of Thera. During a pause of the volcanic activity that occurred on AD September 29, 1650 (OS) (Figure 2.3), a sea swell encircled the whole of Thera Island. The tsunami inundated the eastern coast of the island and swept away churches, enclosures, boats, trees, and cultivated land. It was reported that on the East and West coast of Patmos Island, the sea rose 164 ft. (50 m) and 98 ft. (30 m), respectively, on Ios Island the waves rose to 50 ft. while at Sikinos Island the tsunami swept 350 ft. inland (Galanopoulos and Bacon, 1970). Ships and fishing boats moored at Heraklion, Crete, were violently swept offshore, while vessels were crushed when the wave overtopped the ~4 m high city walls. The generation mechanism of that tsunami is still puzzling (Dominey-Howes et al., 2000a; Nomikou et al., 2012) and, therefore, it is discussed further in the next section. Geological evidence of this tsunami is reviewed in the next section of this chapter.

A careful examination of the documentary sources available for the eruptive activity of Columbo during AD 1650 revealed that on November 4, 1650 (OS) a second but smaller tsunami was also observed in Santorini (Triantafyllou et al., 2015).

The most recent large Mediterranean tsunami occurred in Cyclades, South Aegean Sea, on AD July 9, 1956 after a M_w 7.5 crustal earthquake (Figures 2.2 and 2.3) (Galanopoulos, 1957a,b; Ambraseys, 1960; Papazachos et al., 1985) and was associated with normal faulting (Papadopoulos and Pavlides, 1992; Perissoratis and Papadopoulos, 1999;

Figure 2.4 ***Fishing boat ashore in Kalymnos Island due to the strong tsunami of July 9, 1956 (Galanopoulos, 1957b).***

Okal et al., 2009, and references therein). The tsunamigenic source, trending North East-South West, was up to 100 km in length (Papadopoulos and Pavlides, 1992). Initial estimates of the near source wave height varied between 15 m and 30 m in Amorgos and Astypalaea (Galanopoulos, 1957; Ambraseys, 1960). Four people were drowned while extensive destruction was noted in port facilities, small and large vessels, cultivated land and property (Figures 2.4–2.6). Marine-deposited sediments were described in Astypalaea (Dominey-Howes, 1996; Dominey-Howes et al., 2000). The tsunami was recorded by near-field tide gauges in Laki, Leros Island and as far as in Souda, Crete (Galanopoulos, 1957) and Yafo, Israel (Goldsmith and Gilboa, 1985) with wave amplitudes of ~4 m, 1.5 m, and 28 cm, respectively.

In the east Aegean Sea, a distinct tsunami-prone area is associated with earthquakes occurring near Chios Island, where local tsunamis of intensity 3 or 4 were observed on AD March 20, 1389; May 12, 1852; September 8, 1852; November 13, 1856 (Figure 2.7); February 2, 1866; April 3, 1881; and July 23, 1949. In the North Aegean Sea, according to the Greek historian Herodotus (484–426 BC), a destructive sea wave reportedly struck Potidaea in Chalkidiki Peninsula, in 479 BC, April. The description may imply tsunami action but no earthquake was reported. Reicherter et al. (2010) described possible tsunami sediment deposits in Thermaikos Gulf that could be attributed to that historical event. Remarkable local tsunamis of intensity 3 were observed after strong earthquakes in Samothraki Island on AD February 9, 1893 and in Ierissos, Chalkidiki peninsula, on AD September

Figure 2.5 ***Sea retreat in Patmos Island due to the strong tsunami of July 9, 1956 (Galanopoulos, 1957b).***

Figure 2.6 ***The July 9, 1956 tsunami inundation in Patmos island (Galanopoulos, 1957b).***

26, 1932 (Figure 2.7). In AD 544 an earthquake-induced destructive inundation hit the coast of Thrace facing the North Aegean Sea. However, its origin remains questionable since strangely enough it was reported that it inundated also the Bulgarian coast of the Black Sea (Papadopoulos et al., 2011, and references therein).

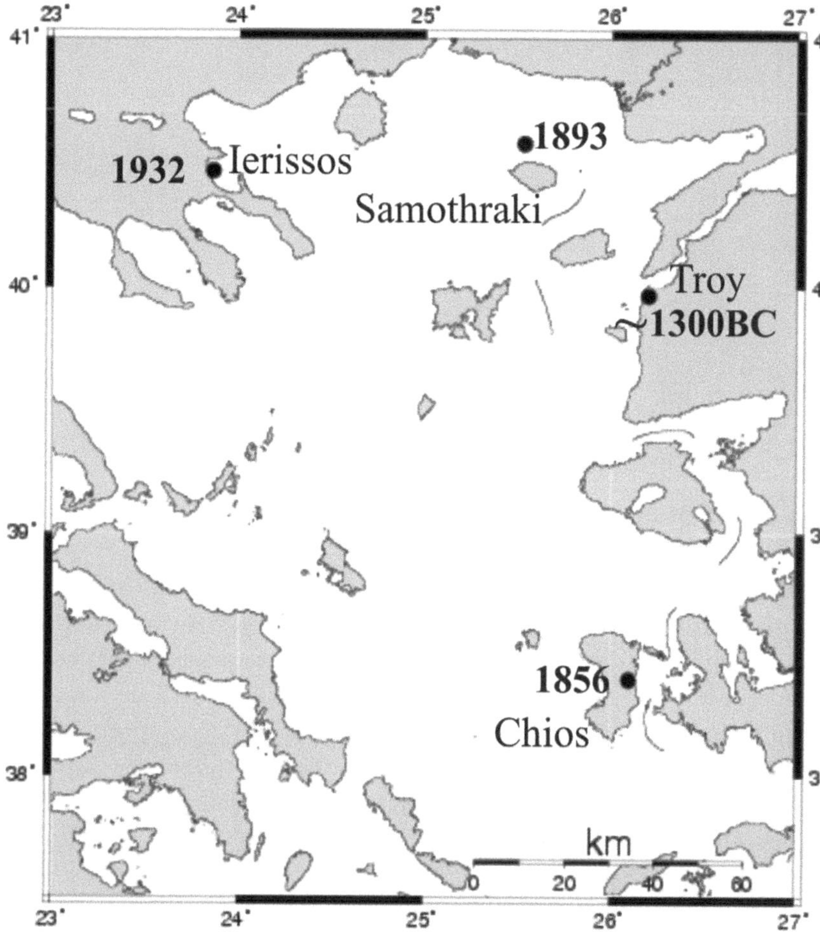

Figure 2.7 ***Important tsunamis in the east and north Aegean Sea.***

2.2.1.3 Tectonic Rifts of Corinth and Evoikos Gulfs, Central Greece

The rift of Corinth Gulf, Central Greece, is highly prone to tsunamis (Figure 2.8). Regardless the tsunami size, this rift structure is characterized by the highest rate of tsunami occurrence in the European–Mediterranean region. This is explainable by the concurrence of several factors favoring the tsunami generation, such as high seismicity, susceptibility to coastal/submarine landsliding and steep bathymetry (Papadopoulos, 2003a). The tsunami rate is higher in the west side of the Gulf, which is consistent with the fact that the seismicity rate is also higher than on the east side. However, the Corinth Gulf tsunamis are not the largest in the Mediterranean region and are not able to propagate outside the gulf. A long number of documentary

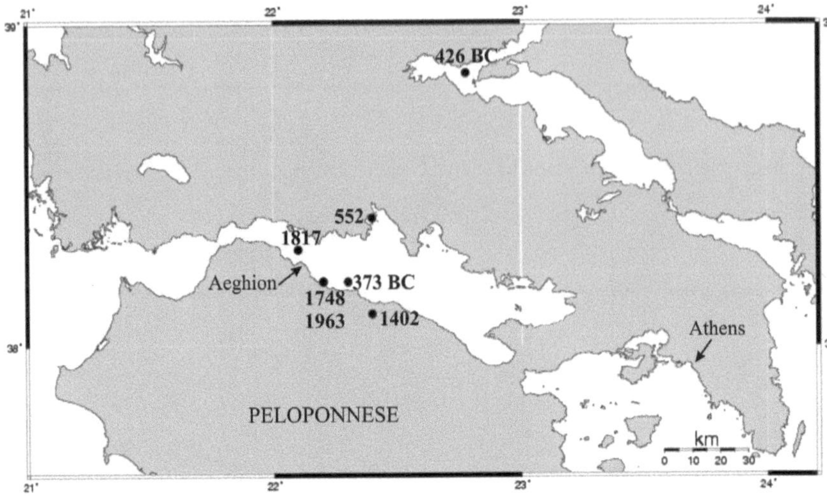

Figure 2.8 ***Important tsunamis in Corinth Gulf and Maliakos Gulf.***

sources (Guidoboni et al., 1994), support that in 373 BC the coastal town of Helike, located about 7 km east of the modern city of Aeghion in South West Corinth Gulf, was destroyed by a strong earthquake and its associated local tsunami. Ten Spartan ships, which were at anchor, were destroyed by the wave. Papadopoulos (1998) suggested a scenario comprising three sequential geological events: (a) strong earthquake, (b) extensive liquefaction of an extensive coastal strip, and (c) tsunami inundation. The author estimated earthquake magnitude of at least 6.6, which is about equal to the magnitude assigned to a quite similar event that happened in the same area on AD December 26, 1861 (Schmidt, 1879).

Strong, lethal earthquakes occurring on AD May 25, 1748 and AD August 23, 1817 generated local but still powerful tsunamis causing human losses and extensive damage in the coastal zone of Aeghion, West Corinth Gulf (Papadopoulos, 2003a and references therein). These two tsunamis were quite similar in their hydrodynamic features and their consequences. The AD June 1402 tsunami occurred further east in the south coast of the Corinth Gulf (Figure 2.2). It was very powerful and followed a large, possibly near shore large earthquake having its source close to the town of Xylokastro (Figure 2.8) (Evagelatou-Notara, 1987/1988; Papadopoulos, 2003a).

Seismically triggered landslides caused local tsunamis along the North coast of Corinth Gulf on AD June 11, 1794; July 6, 1965; February 11, 1984; and June 15, 1995. An aseismic, damaging tsunami generated by a sediment slump at a river mouth hit both coasts of western Corinth Gulf killing two people on AD February 7, 1963 (Galanopoulos et al., 1964). This tsunami is further examined later given that it is one of the key events for understanding the tsunami generation from aseismic landslides. A similar but less intense wave was observed near Aeghion on AD January 1, 1996 (Papadopoulos, 2003a).

About 90 km to the northeast from Corinth Gulf, the North West–South East trending tectonic rift of Evoikos Gulf, including its northwestern termination called Maliakos Bay, is remarkably active from the seismicity point of view. However, only a few tsunamis have been historically reported there. The historian Thucydides (460–403? BC) described a strong earthquake occurring in 426 BC (Figures 2.2 and 2.8). Later on, geographer Strabo (64 BC–19 AD) reported that a tsunami, which associated the earthquake, violently inundated coastal localities of Maliakos Bay (Antonopoulos, 1992b). Sediment layers of likely tsunamigenic origin were described by Gaki-Papanastassiou et al. (2001) but this preliminary evidence was not further supported. On the other hand, Papaioannou et al. (2004) evaluated additional historical and archeological reports and suggested that the 426 BC seismic event was rather moderate, arguing that the large tsunami from that period may have occurred during the third century BC, and that previous researchers amalgamated the two events into the earlier one at 426 BC.

The Byzantine historian Procopius reported on earthquakes that struck Corinth Gulf settlements in AD 552 (Ambraseys, 2009, and references therein). Procopius described also a strong tsunami in the Maliakos Bay, which, however, looks like the one of 426 BC as described by classic authors. Therefore, one may suggest that very likely Procopius did not describe a tsunami caused by the AD 552 earthquake but just reproduced classic sources regarding the 426 BC tsunami. Scholars have noticed (Morrisson, 2007) that Procopius was prone to imitate classic Greek historians.

2.2.1.4 Cyprean Arc and Levantine Sea

The area of Cyprus and Levantine Sea (Figure 2.9) is of relatively low tsunami frequency (Fokaefs and Papadopoulos, 2007). Archeological excavations in Kourion, South West Cyprus, elevated at about 100 m above the sea level, revealed a destruction horizon initially attributed to the AD July 21, 365 earthquake and tsunami (Soren, 1988). Although the archeological findings revealed an impressive seismic destruction layer containing human and animal skeletons (Figure 2.10), the association of the earthquake with a tsunami is only speculative since no tsunami evidence was found at all. In fact, it is highly unlikely that a tsunami could reach at that high elevation

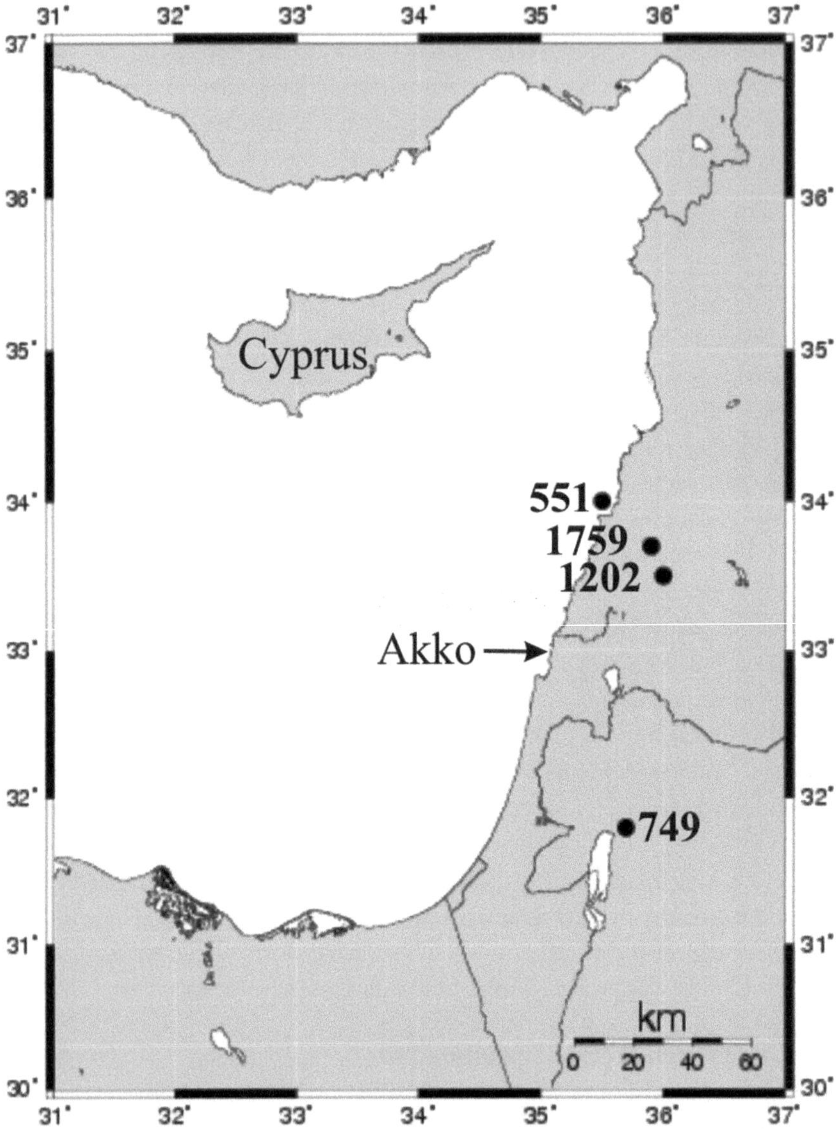

Figure 2.9 ***Important tsunamis in the Cyprus-Levantine Sea area.***

(Figure 2.11). Besides, as analyzed earlier, the large AD 365 earthquake at all evidence ruptured the western Hellenic Arc. Consequently, an earthquake source situated to the west of Crete Island does not account for destruction caused as far as South West Cyprus at an epicentral distance of about 900 km.

Figure 2.10 ***The part of archeological site of Kourion, South West Cyprus, where excavations revealed human and animal skeletons buried in ruins (Soren, 1988) as a result of the strong earthquake of c. AD 370.*** *(Photo courtesy by G.A. Papadopoulos.)*

Figure 2.11 ***The west coastal segment of Kourion as it is seen from the ancient town.*** *(Photo courtesy by G.A. Papadopoulos.)*

From historiographic analysis and coin dating, it comes out that very likely Kourion was hit by a local earthquake occurring without tsunami around AD 370 (Guidoboni et al., 1994; Ambraseys, 2009). In South East Cyprus a damaging tsunami of intensity 7 (Fokaefs and Papadopoulos, 2007) was reported in association to the very large (M = 7.6, Guidoboni and Comastri, 2005) earthquake of AD May 22, 1202, which ruptured the Dead Sea Fault Zone (Salamon et al., 2009). The tsunami was also reported from the Syrian coast. Submarine landslide likely triggered by a possibly inland earthquake might be the tsunami-generating mechanism. On AD May 11, 1222 (Figure 2.12) and on September 10, 1953 two strong earthquakes occurring in South West Cyprus caused also local tsunamis assigning intensity of 5 and 3, respectively (Fokaefs and Papadopoulos, 2007).

In the left-lateral strike-slip Levantine rift, tsunami-generating earthquakes have been identified from documentary sources (Figure 2.9) (see reviews in Amiran et al., 1994; Guidoboni et al., 1994; Darawcheh et al., 2000; Elias et al., 2007; Salamon et al., 2007; Ambraseys, 2009). On AD July 9, 551 the sea retreated for a mile and many ships were destroyed along the coasts of Phoenicia, modern Lebanon, particularly in the cities of Beirut and Tripoli, as well as in Syria and Palestine. This tsunami was due to a large submarine earthquake (Figure 2.12) very likely caused by thrust faulting (Elias et al., 2007). After an earthquake on AD January 18, 746 or 749, which affected Palestine, Jordan, and Syria, the waves "rose up to the sky" and destroyed cities and villages. This is one of the historical events, which certainly deserves further examination for better understanding the tsunami source. Tsunamis were reported after the strong earthquakes on AD December 5, 1033, that shook the region around the Jordan Valley, on AD May 29, 1068, an event possibly centering at the south of the Levantine rift, and on AD January 14, 1546 near the Dead Sea (Ambraseys, 2009).

The two millennia of historic time frame are much shorter than the repeat time of large earthquakes in many of the tectonic structures as well as of sedimentary slumps upload in the region of Levantine Sea. Given the limited time span of historical documentation and the stabilization of the sea in its present level only a few thousands of years ago, the evidence of tsunamis that have occurred before that time is hard to trace and difficult to find, if there at all (Papadopoulos et al., 2014a). It means that potential tsunamigenic sources, the repeat time of which is longer than $\sim$5 ka, should also be considered for hazard evaluation, even if there is no historical evidence for their activity. The Nile Cone is such an example (Maldonado and Stanley, 1976; Salamon, 2011). The only evidence of a tsunami that might have

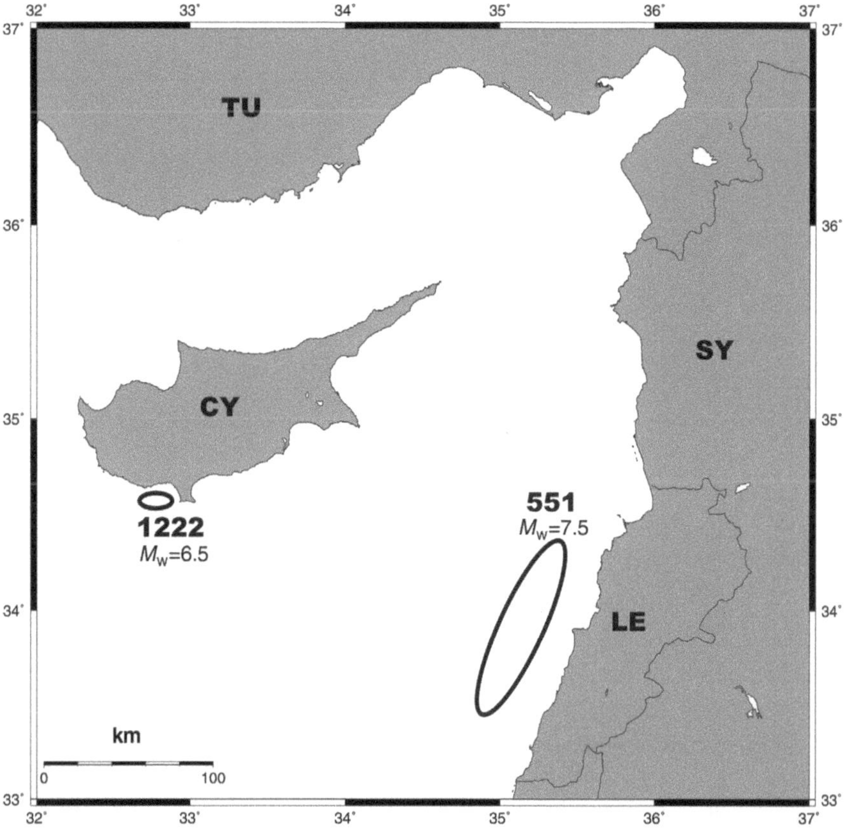

Figure 2.12 ***Rupture zones of strong tsunamigenic tsunamis in Levantine Sea and Cyprus (Papadopoulos and Papageorgiou, 2014).***

been originated from there is the 20 BC orphan tsunami between Alexandria and Pellusium, Egypt (see Salamon et al., 2009, and references therein). However, the tsunami record in the eastern Mediterranean motivated the performance of tsunami hazard studies in geographical spots of particular economic interest (Finkl et al., 2012).

2.2.2 Marmara Sea and Black Sea

In the Marmara Sea, the most tsunami-prone side is the east one due to its higher seismicity with respect to the west side. Intensity 3 or 4 tsunamis were caused by earthquakes on AD January 26, 447; October 26, 740; September 2, 1754 and April 19, 1878 in the Izmit Bay, on AD September 25, 478; August 10, 1265; October 18, 1343 (Figure 2.2); May 25, 1419; September 10, 1509; and July 10, 1894 in Constantinople (Istanbul), on

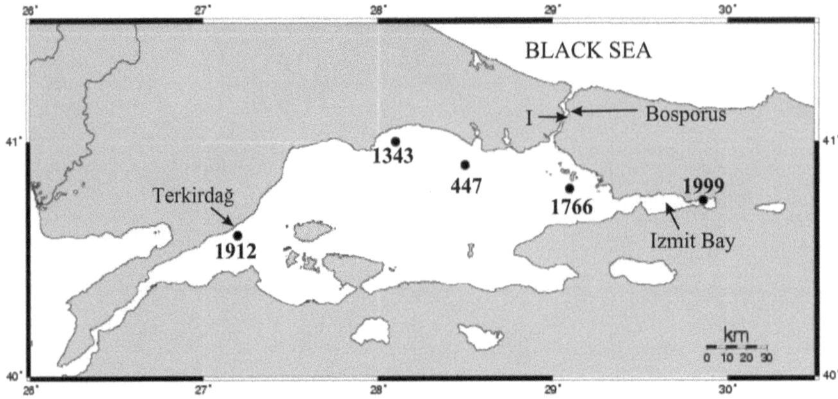

Figure 2.13 ***Important tsunamis in Marmara Sea.***

AD May 22, 1766 in Bosporus Straits, and on August 9, 1912 in Terkirdağ (Altinok et al., 2011) (Figure 2.13). The large M_w 7.4 Izmit earthquake of August 17, 1999, caused by right-lateral strike-slip faulting with significant normal component, generated damaging waves up to 2.5 m high in Değirmendere, South coast, and elsewhere in the Izmit Bay (Altinok et al., 1999; Yalciner et al., 2002; Tinti et al., 2006). The local tsunami was of intensity up to 4 (Papadopoulos, 2001).

Data on the Black Sea tsunamis were published by several authors (Pelinovsky, 1999; Altinok and Ersoy, 2000; Yalciner et al., 2004). An updated data compilation and critical evaluation of tsunamis occurring in the Black Sea and the Azov Sea from antiquity up to the present have indicated that the tsunami hazard is very low but still not negligible (see Papadopoulos et al., 2011, and references therein). Most of the tsunamigenic seismic sources are concentrated in three main zones, which are: (a) around the peninsula of Crimea; (b) offshore Bulgarian coast; and (c) in the eastern side of the Black Sea.

From Byzantine historical sources, it appears that a strong sea wave, very likely a tsunami, that hit coastal settlements of Bulgaria, was caused by a very strong earthquake occurring on AD 544 or 545 offshore modern Varna (Figure 2.2). Ranguelov et al. (2008a) presented field evidence from an archeological site in Balchik city, situated ~20 km north of Varna. Namely, the Cybele Temple was affected by fire and roof collapse, very possibly due to that earthquake. The floor was flooded by seawater, which left behind a layer of sand and shells. The Temple is of the Hellenistic period and located about 180 m inland from the present shoreline (Figure 2.14).

Figure 2.14 ***View of the archeological site of Cybele near Balchik, Bulgarian Black Sea coastal segment, where evidence of tsunami deposits were found (Ranguelov et al., 2008a).*** *(Photo courtesy by G.A. Papadopoulos.)*

According to Ranguelov et al. (2008b), an aseismic tsunami-like sea disturbance was observed on May 7, 2007 on the Bulgarian Black Sea coastal segment of about 150 km in length and centering near Balchik. The chief period of the oscillations was 4–8 min while slight damage was caused. Although the hypothesis of tsunami generation from submarine landslides should not be ruled out, Vilibić et al. (2010) analyzed the tsunami features and atmospheric conditions and concluded in favor of a meteotsunami, the first of such kind reported in the Black Sea.

2.2.3 Tyrrhenian Sea, Calabrian Arc, and Adriatic Sea

The Tyrrhenian Sea has also been the source of tsunamis in the historical past. In Tuscany, a tsunami assigned intensity 4 was reported to occur on AD March 5, 1823. In the Aeolian Islands, tsunamis assigning intensity 3 or 4 produced by landslides in Stromboli volcano, were historically documented on AD July 3, 1916; May 22, 1919; September 11, 1930; and August 20, 1944 (Maramai et al., 2005a) (Figure 2.15). The best studied case was the one that took place on December 30, 2002 during an eruptive episode in Stromboli (Tinti et al., 2005a,b) (Figures 2.16 and 2.17).

Another tsunamigenic zone is that of eastern Sicily and Messina Straits. An extreme event occurred in Tyrrhenian Calabria on AD February 6, 1783 (Tinti and Guidoboni, 1988) since a huge earthquake-induced rockfall

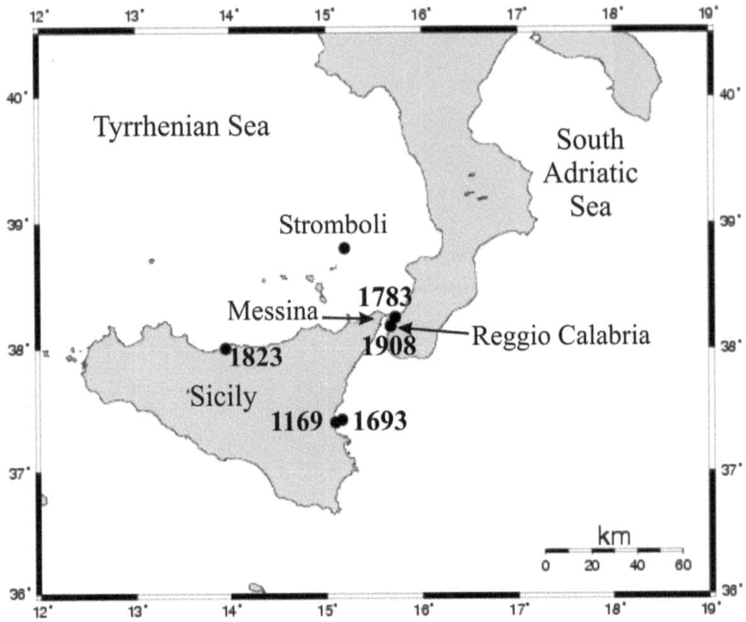

Figure 2.15 *Important tsunamis in South Italy.*

Figure 2.16 ***Outdoor damage caused in Ficogrande, Stromboli, by the December 30, 2002 tsunami.*** *(Photo courtesy by G.A. Papadopoulos.)*

(Bozzano et al., 2011) triggered a tsunami of intensity 6. Inundation heights of 6–9 m were observed and more than 1500 lives were lost. In North East Sicily the AD January 11, 1693 devastating earthquake, that claimed about 70,000 victims, caused an intensity 7 tsunami.

Figure 2.17 ***Indoor damage caused in Ficogrande, Stromboli, by the December 30, 2002 tsunami.*** *(Photo courtesy by G.A. Papadopoulos.)*

One of the most lethal earthquakes ever reported in Italy was the one of AD December 28, 1908 measuring magnitude M_w 7.1 (Figure 2.15). Major towns in southern Italy, including Messina and Reggio Calabria, were completely destroyed with a death toll of ~60,000. After the devastating earthquake, a violent tsunami hit the Messina Straits (Baratta, 1910; Omori, 1910; Platania, 1909). At least three large waves were observed causing many deaths and severe damage to ships, buildings, and property. Tsunami wave heights of 13 m and 11.7 m were observed along the Calabrian coast at Pellaro and on the Sicilian coast at S. Alessio, respectively (Tinti and Maramai, 1996), where a tsunami of intensity 10 was assigned (Papadopoulos, 2009). In earlier times, a tsunami likely similar to that of 1908 was reported to hit Messina in AD 1169.

In the north side of the Adriatic Sea small-to-moderate tsunamis were historically reported (Maramai et al., 2007; Paulatto et al., 2007; Pasarić et al., 2012). On the Italian coast, tsunami source is controlled by the seismicity of the Gargano promontory (Tinti et al., 1995). The destructive earthquake of AD July 30, 1627, which may have been associated with the Apricena normal fault on land (Patacca and Scandone, 2004), caused a powerful tsunami (Guidoboni and Tinti, 1988) with assigned intensity 6 (Papadopoulos, 2009). On the eastern side of the Adriatic Sea moderate-to-strong tsunamis were reported in Albania in AD 1866 with assigned intensity 7 (Soloviev et al., 2000; Papadopoulos, 2009, and references therein).

2.2.4 Western Mediterranean Basin

In the basin of western Mediterranean tsunamis were reported (Figure 2.18) in North Algeria to have been caused by strong earthquakes occurring on AD January 2, 1365; May 6, 1773; August 21–22, 1856; September 9, 1954; and October 10, 1980. Those tsunamis were of estimated intensities up to 4 (Papadopoulos, 2009). Soloviev et al. (2000) suggested that the location of most of these earthquakes on land in North Algeria triggered submarine landslides that generated powerful turbidity currents. However, the May 21, 2003 small-to-moderate tsunami produced by the M_w 6.8 Boumerdes–Zemmouri (Algeria) earthquake in the continental margin of Algeria was attributed to a thrust coseismic faulting (Yiga, 2003; Meghraoui et al., 2004; Wang and Liu, 2005; Alasset et al., 2006).

In the Ligurian Sea and the Côte d'Azur, an aseismic, submarine slope failure that occurred on October 16, 1979 during the construction of the new Nice airport produced a destructive tsunami wave of 3 m in height that was observed as far as Antibes about 15 km to the west of Nice. The AD July 20, 1564 and AD February 23, 1887 earthquakes (Eva and Rabinovich, 1997; Larroque et al., 2012) triggered tsunamis inundating the coast from Nice to Antibes and from Genoa to Cannes, respectively.

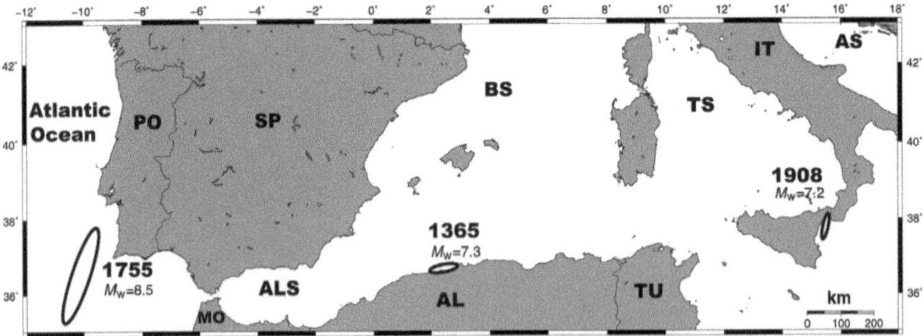

Figure 2.18 ***Rupture zones of strong tsunamigenic tsunamis in western Mediterranean Sea and in South West Iberia (Papadopoulos and Papageorgiou, 2014).***

2.2.5 North East Atlantic Ocean and North Sea

2.2.5.1 South West Iberian Margin

Off the coast of South West Iberia, particularly in the external part of the Gulf of Cádiz, there is a highly seismogenic zone producing very strong tsunamis (Baptista and Miranda, 2009). The most famous was the disastrous November 1, 1755 earthquake-tsunami event (Figure 2.18), which affected all countries around the Gulf of Cádiz as well as a large part of the Atlantic Ocean coasts (see reviews in Fonseca, 2005; Oliveira, 2008). Wave heights up to 15 m were described in Cape St.Vincent and along the Gulf of Cádiz. In Lisbon the number of casualties, due exclusively to the tsunami, is estimated to be close to 900 and the run-in in Lisbon downtown is estimated at ~250 m (Baptista, 1998). However, the location and type of source is still debated, while suggestions for an overestimation of both the earthquake and the tsunami were published (Blanc, 2009). The Horseshoe earthquake (M_w 7.9) of February 28, 1969 produced only a small tsunami recorded in several tide gauges with maximum amplitude of 0.6 m in Casablanca (Baptista et al., 1992).

In the Azores and Canary islands small amplitude (less than 1 m) tsunami-like sea disturbances were observed and recorded in tide gauges after some strong earthquakes, such as the ones of May 8, 1939 (M_s 7.1); November 25, 1941 (M_s 8.4); February 28, 1969 (M_s 7.9); May 26, 1975 (M_s 7.9); and January 1, 1980 (M_s 6.7) (Baptista and Miranda, 2009). However, the 1755 tsunami inundated all of the Azores islands, with maximum run-up (11-15 m) in Terceira (Andrade et al., 2006). Borges et al. (2008) reported that in total 29 tsunamis have been historically known in Azores.

2.2.5.2 British Isles and North Sea

In the British Isles, apart from far-field tsunamis reaching their coasts, such as the Lisbon large tsunami of November 1, 1755, a few tsunami-like sea disturbances linked to earthquakes were reported, for example, in the Dover Straits, during historical times (Haslett and Bryant, 2008). In Norway, several tsunamis were historically reported, all of them being attributed to rockfalls and landslides. One of the most important events, of run-up height more than 70 m, was produced by the February 7, 1731 tsunami into the fjord Storfjorden when more than 100,000 m of rocks landslided from the mountain Rammerfjell. Extensive damage and 17 casualties were reported (Jørstad, 1968). Geological evidence about the strong tsunami produced by the large-scale Storegga submarine landslide occurring offshore Norway around 8200 years BP is discussed in the next section.

2.3 GEOLOGICAL AND ARCHEOLOGICAL SIGNATURES

In analogy to the geology of earthquakes, the geology of tsunamis is of particular importance for the identification of geological traces of tsunamis of the past, that is, of paleotsunamis. This is of crucial importance given that tsunamis are rare events and, therefore, the historical tsunami cataloges available for hazard and risk assessment are vastly incomplete. In recent times it became more and more obvious that geological records of past tsunamis are also a precious source of information to integrate and extend back in time historical records (Papadopoulos et al., 2014a).

The basic principle for the geological identification and dating of past tsunamis rests on the observation that tsunami waves mobilize and deliver a large amount of sediments both inland and offshore. Thus, the recognition of tsunami-related deposits can be directly translated into the geological record of a tsunami. Dating of these deposits through common Quaternary geochronological techniques, such as accelerator mass spectrometry (AMS) radiocarbon dating, optically stimulated luminescence (OSL), short-lived radionuclides, tephrochronology, paleomagnetism, etc., yield dates tsunami. In some instances, tsunamis left their impact in archeological sites too, and, therefore, the archeological identification and dating of a tsunami event, often combined with geological evidence, makes a particular case of paleotsunami record.

This section summarizes the main achievements regarding the recognition of paleotsunamis from geological, geomorphological, and archeological field evidence.

2.3.1 Onshore Signature

Paleotsunami studies started about 20 years ago or so providing firm evidence that during inundation of coastal areas, tsunamis are capable of transporting a large amount of sediments inland, ranging from fine-medium sand to boulders and megaclasts. This was verified after recent studies on the onshore geological signatures of the Indian Ocean 2004 and Japan 2011 mega tsunami events (Goto et al., 2011). This sedimentary material is eroded from both the seafloor and the shore by the waves approaching and flooding the coastal zone. As a consequence, tsunami deposits in continental areas are sediments at least partially of marine origin, deposited by a high-energy hydrodynamic process.

The investigation of tsunami deposits is favored in coastal lakes, lagoons, and almost flat fluvial plains, which are characterized by low-energy environment. The direct effects of coastal tsunami inundations are usually reflected in geomorphological features, such as beach erosion, destruction of sand barriers and/or formation of landward washover fans, hummocky topography and large scars (Goff et al., 2009).

2.3.2 Medium-Fine Grained Deposits

Studies on modern- and paleo-tsunami deposits have revealed that they are characterized by some general features. Namely, the average grain size of the deposits decreases landward and upward. This depends on the nature of the available near- and onshore sediments (Goff et al., 2012) as well as on the hydrodynamic conditions during transport and sedimentation.

The tsunami deposit thickness also varies depending on the energy of waves and type of deposits involved as well as on the local topography. In general, the thickness of tsunami deposits does not exceed a few tens of centimeters (Morton et al., 2007). Usually in tsunami deposits the number of layers is limited to only a few and they do not display peculiar sedimentary structures apart from rare laminae. On the other hand, the basal contact is abrupt. Biological indicators of the marine origin of a specific layer may include diatoms (Nichol et al., 2007), ostracods (Ruiz et al., 2010) and foraminifera (Mamo et al., 2009) although no "characteristic" species are associated with tsunami deposits.

2.3.3 The Stratigraphic Record

2.3.3.1 Mediterranean and Marmara Sea

In the Mediterranean region of particular value are studies aiming to verify geological evidence by archeological findings and/or historical tsunami documentation and integrate them into a unified picture of the paleotsunami history. An important case regards the sediments deposited by the LBA large tsunami produced by the great eruption of Thera (Santorini). Such sediments were observed in South West Turkey and North Crete (Minoura et al., 2000), in Thera (McCoy and Heiken, 2000), and again in North Crete (McCoy and Papadopoulos, 2001; Bruins et al., 2008) (Figures 2.19 and 2.20).

Another important example is the excavated harbor of Phalasarna in North West Crete. The classical and Hellenistic town of Phalasarna (Figure 2.21) was mentioned in the texts of several ancient geographers, with particular note taken of its closed harbor, measuring 100 m × 75 m in dimension (Hadjidaki, 1988). According to Hadjidaki (2001), the archeologist who excavated and revealed Phalasarna, this coastal town was put out of business during the first century BC. The most plausible reason is that Roman naval forces destroyed Phalasarna in 68–66 BC. Should this

Figure 2.19 ***During excavation works in the archeological site of Gouves (Vallianou, 1996), located about 15 km to the east of Knossos, north coast of Crete, it was observed that the floor of the late Minoan Potters' workshop is covered by a thin veneer of carbonate sand and an overlying 10–20-cm-thick pumice layer.*** The excavated site is situated 30–90 m inland from the Minoan harbor installation and 2–3 m high above the present sea level. The carbonate sand, composed of unsorted grains of skeletal fragments, is marine in origin and it is interpreted that the sand layer was deposited during seawater flooding in late Minoan time very likely by the Late Bronze Age Thera tsunami (Minoura et al., 2000). *(Photo courtesy by G.A. Papadopoulos.)*

hypothesis be correct, then the harbor was already out of order when a historically documented earthquake and tsunami occurred in AD 62 (or 66) in the area (see review in Papadopoulos, 2011). The two long canals that connected the Phalasarna harbor to the sea are both situated today on dry land, as is the port itself, possibly because of the abrupt uplift during the mega earthquake of AD 365 (Hadjidaki, 1988, 2001; Frost and Hadjidaki, 1990).

Tsunami sediment deposits were described in Phalasarna but this is still controversial. Pirazzoli et al. (1992) were able to identify two layers of "coarser material with blocks" in two out of three trenches and interpreted that both represent sediment layers deposited by the AD 66 and AD 365 tsunamis. On the other hand, Dominey-Howes et al. (1998) reported that the foraminiferal assemblage indicates tsunami deposition around AD 66, but that there is no bio- or lithostratigraphic evidence to infer sedimentary deposition associated with the AD 365 tsunami. They found it hard to explain why the stratigraphy at Phalasarna records no evidence of such a large displacement. One viable explanation for the relatively limited effects of the AD 365 tsunami in the sediment stratigraphy of Phalasarna is that the site was already uplifted by 6.6 m, only a few minutes before the wave

Figure 2.20 ***Palaeotsunami trenching at distance of about 60 m from the seashore near Didim, south west Turkey, revealed a sediment deposit attributed to the LBA Thera tsunami (Minoura et al., 2000).*** In this part of the section, the dark tsunami deposit is layered at depth of about 25 cm from the ground surface and is overtopped by volcanic tephra (white layer) of the LBA eruption. *(Photo courtesy by G.A. Papadopoulos.)*

Figure 2.21 ***View of the excavatation by Dr E. Chatzidaki of the archeological site of Phalasarna, north west Crete.*** Part of the docks of the Hellenistic harbor are shown at the left-hand side. *(Photo courtesy by G.A. Papadopoulos).*

arrived (Pirazzoli et al., 1992). In a vibrocore tsunamigenic sand layers were found in Cape Punta, South East Peloponnese (Scheffers et al., 2008). A wood fragment found at the base of tsunami deposits was dated to ~250 cal AD, which may be a signature of the AD 365 large tsunami.

Sedimentological analyses combined with geomorphological and archeological findings in the Ambrakian Gulf, in Akarnania as well as in Lake Voulkaria, all in coastal zones of Ionian Sea, North West Greece, Vött et al. (2007, 2008, 2009, 2011) supported that were able to identify several sediment layers attributed to a series of prehistorical and historical tsunami events. However, the geological record is not supported by historical tsunami documentation, with the possible exception of the AD 365 event.

Along the coast of eastern Sicily, Italy, through a multitheme approach including detailed geomorphologic and geologic surveys, satellite images, and aerial-photos interpretation, coring campaigns, and laboratory analyses, De Martini et al. (2010) found evidence for a record of 6–7 marine inundations, which occurred during the past 4 ka. Peculiar grain size characteristics and foraminiferal assemblages of the studied layers, their distance from the shoreline, their ages and frequency with respect to historical cataloges and previous studies, were used to identify tsunamis as the most likely process of emplacement, namely the AD 1693 local one, and the distant events of AD 365 in Crete and of LBA in Thera.

In Dalaman, South West Turkey, a series of three sea sand layers (Figure 2.22) were discovered and correlated with three historically documented tsunamis occurring in Rhodes Island (South East Greece) in AD 1303, 1481, and 1741 (Papadopoulos et al., 2005b, 2012a). In the Corinth Gulf, continental Greece, several layers of sediment deposits were described in two test sites and were attributed to prehistorical and historical tsunamis including the powerful, destructive waves of AD 1402 and 1817 (Kontopoulos and Avramidis, 2003; Kortekaas et al., 2003, 2011).

In Saros Gulf, North East Aegean Sea, dune sand containing abundant pumice drifted inland was interpreted as the result of tsunami action (Erginal et al., 2009). However, such an interpretation is not strongly convincing. OSL dating showed that the event occurred at least 670 years ago. Therefore, Erginal et al. (2009) tentatively correlated the hypothetical tsunami with a strong earthquake, which occurred in the southeastern Aegean Sea around AD 1672. But the occurrence of such an earthquake event is of low reliability (Papadopoulos et al., 2007a). Besides, it is unlikely that an earthquake occurring in South East Aegean produced a tsunami that propagated well enough all the way of the complex island physiography of the Aegean Sea and arrived to cause impact in the North East Aegean.

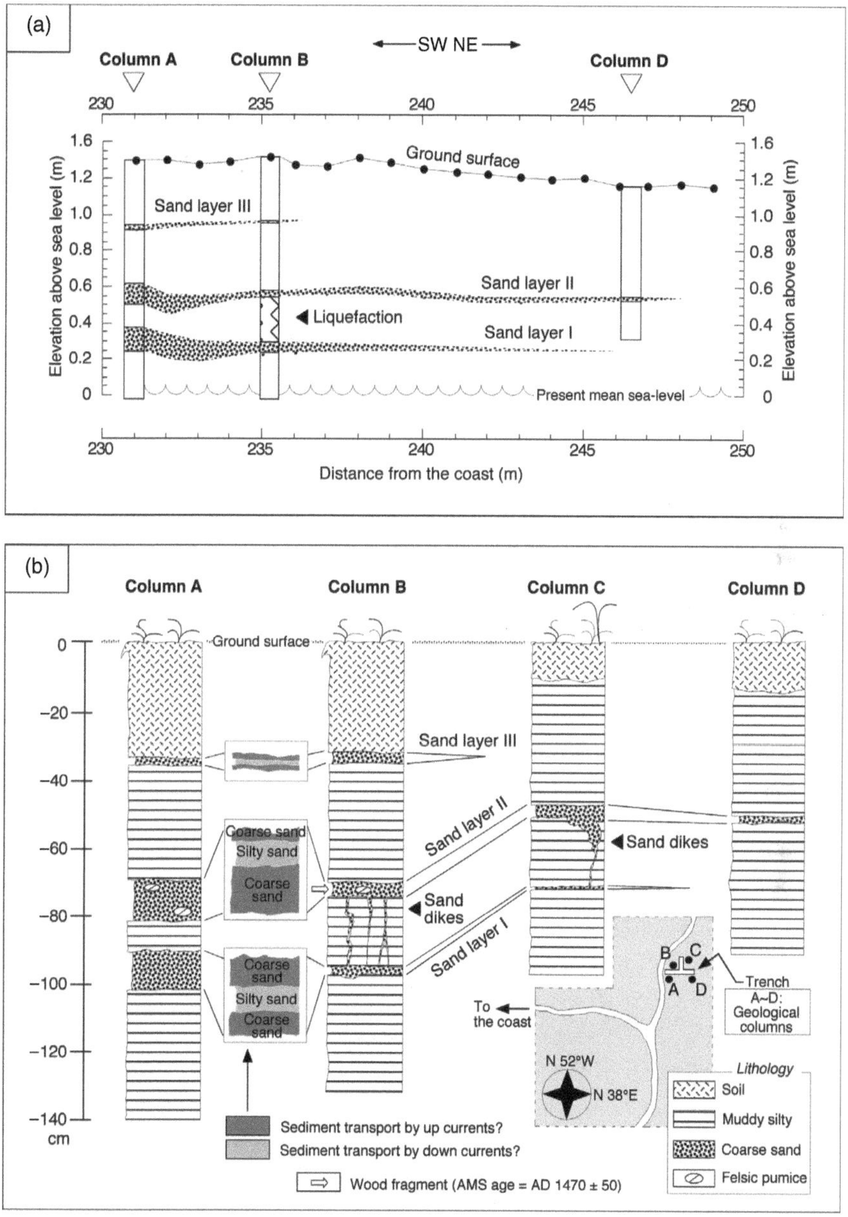

Figure 2.22 ***Stratigraphic section of the palaeotsunami trenching in Dalaman, south west Turkey (after Papadopoulos et al., 2012a).*** (a) General view, and (b) detailed view. Three sediment layers of sea sand were attributed to the 1303, 1481, and 1741 historically documented tsunamis (see text). Sand dykes directed upwards from layer I to layer II are attributed to liquefaction of sand layer I due to the large 1481 earthquake that caused the tsunami layer II.

Figure 2.23 ***The early Byzantine archeological site of Perissa, east coast of Santorini Island, in the sections of which two sand layers, very likely of tsunami origin, were found at the spot where the author stands.*** See closer view in Figure 2.24. *(Photo courtesy by I. Triantafyllou.)*

A very likely tsunamigenic sediment layer, possibly dated to the eleventh century AD, was discovered in fluvio-alluvial sequences on the northern coast of the Marmara Sea (Minoura et al., 2005). The layer consists of unsorted silty coarse sand including terrestrial molluscs and charcoal fragments. The AMS radiocarbon ages of the shells were estimated at around 400 BC and AD 300, 400, and 1000. A landslide-associated tsunami that occurred in the Marmara Sea in the middle of eleventh century AD and invaded the fluvial plains was proposed by Minoura et al. (2005).

In Santorini (Thera) island, the strong tsunami that was caused by the September 29, 1650 (OS) extra-caldera submarine eruption of Columbo (Papadopoulos, 2009) and was historically documented to hit violently the eastern coastal zone of the island (Dominey-Howes et al., 2000a and references therein), has been identified in two coastal spots at the eastern coast. A sediment deposit consisting of sea sand found in an archeological section situated at distance of ~90 m inland and at elevation of ~4 m a.s.l. in Kamari, east coast of Santorini, was interpreted as a geological record of the AD 1650 tsunami (Papadopoulos, 2009).

Very recently, in an archeological section in Perissa, east coast of Santorini, our team of the National Observatory of Athens found two distinct sea sand layers of thickness ranging from 5 cm to 10 cm and interbedded in sediments consisting of Minoan pumice and tephra (Figures 2.23

Figure 2.24 ***Closer view of the section of Figure 2.23 where sand layers were found at depths of 50–60 cm and of 85–90 cm.*** *(Photo courtesy by I. Triantafyllou.)*

and 2.24) transported and deposited in Perissa as the result of erosion action in the nearby limestone Mesa Vouno mountain lying just to the northeast of Perissa. The archeological site is characterized by the presence of graves of early Byzantine era (seventh–eighth century AD). After that era the site gradually covered by the sediments coming from Mesa Vouno and its existence remained unknown until the soil was eroded by the large tsunami of AD September 29, 1650, caused by the Columbo eruption. The site is situated at a distance of about 150 m–180 m from the present shoreline and at elevation of ca. 6 m. There is no doubt that the two sand layers postdate eighth century AD given that they overtop the ground level of the site with the early Byzantine graves.

In historical times, three large tsunamis hit the South Aegean area, the ones of August 8, 1303 in the eastern Hellenic Arc between Crete and Rhodes, the

September 29, 1650 Columbos tsunami and the July 9, 1956 tsunami generated by a large (M_s 7.4) tectonic earthquake that ruptured the submarine trough between the islands of Santorini, Amorgos, and Astypalea. The author of this book visited the area several times in the past and interviewed local people, who testified that in the coastal segment of Perissa the 1956 tsunami did not penetrate a long distance inland, say no more than about 15 m. This is consistent with that in the several reports published about the 1956 tsunami, in which no account was found that the tsunami inundated at a distance that may conclude with tsunami sediment deposition in the Perissa archeological site. Therefore, if the two sand layers were attributed to tsunami inundations the only plausible hypothesis is that they represent the AD 1303 and 1650 tsunamis. On the other hand, extreme sea level due to storm surges may not account for sediment deposition at such long distance and elevation from the sea level as one may conclude from the analysis by Tsimplis and Blackman (1997).

As regards the 1650 tsunami, the historical documentation is clear enough that the wave penetrated inland for a long distance causing great destruction and that certainly caused soil erosion in the archeological site where our investigation was performed. The large 1303 tsunami inundated and caused damage not only in Alexandria but also in the coastal zone of Heraklion where it produced heavy damage and killed people. In the coastal zone of Southern Peloponnese, Scheffers et al. (2008) observed many boulders containing attached marine organisms, dated by AMS ^{14}C around 1300 cal AD, which proved that they were transported from the foreshore environment against gravity by extreme wave events, probably by the large tsunami of AD 1303 produced in the eastern segment of the Hellenic Arc. Also, in Dalaman, South West Turkey, one of the three tsunami sediment deposits found had been attributed to the 1303 wave (Papadopoulos et al., 2005b, 2012a) (Figure 2.22). Although the inundation of the 1303 tsunami in Santorini has not been historically documented, the large size of the tsunami is a good reason to believe that this tsunami arrived and inundated the coastal zone of eastern Santorini a long distance inland.

Although the laboratory examination of the two sand layers found in Perissa is still ongoing, I do conclude that the evidence collected so far favors the hypothesis that the two layers represent onshore deposition caused by the 1303 and 1650 large tsunamis.

Along the Spanish Mediterranean coast, shallow drilling in the lagoon of the Cabo de Gata area provided sedimentary evidence for a paleotsunami (Reicherter and Becker-Heidmann, 2009). The main evidence is that coarse-grained intervals form fining-up and thinning-up sequences that are interpreted as tsunamites. Radiodating indicated that these layers can be ascribed to deposition during the AD 1522 Almería lethal earthquake of estimated magnitude $M > 6.5$.

2.3.3.2 North East Atlantic, North Sea, Baltic Sea

In the South West Iberia area, high-energy, extreme events were identified from sedimentary, paleontological, and geomorphological records on estuaries, marshes, beach-barriers (spit-bars), and some coastal lowlands (see Lario et al., 2010, and references therein). Most of the events were interpreted as tsunamigenic in origin. Lario et al. (2011) summarized results regarding turbidite deposits found offshore South West Iberia. By comparing datasets from both onshore and offshore records, and taking into account some historical and archeological evidence, those authors concluded that at least five tsunami events generated by strong earthquakes have affected the area during the last 7000 years or so before the most recent November 1, 1755 Lisbon earthquake and tsunami. Then, Lario et al. (2011) estimated that in South West Iberia, extreme wave events have a periodicity of 1200–1500 years, although not all were tsunamigenic. The good correlation between the ages of tsunami deposits and widespread deep-sea turbidites strongly suggests a seismogenic origin for the large tsunami events of the South West Iberian Margin (Gràcia et al., 2010).

In relation to the large Lisbon tsunami of 1755 generated in the Atlantic Ocean off the shore of Portugal, tsunami sediment deposits found in Boca do Rio (Algarve) were described as consisting of laterally continuous sand layers, chaotic pebble horizons, large amounts of gravel-sized shell debris and distinctive assemblages of marine microfossils (Dawson et al., 1995; Hindson et al., 1996).

In prehistoric times, the large-scale Storegga submarine landslide taking place on the Norwegian continental slope around 8200 calendar years BP is believed to have caused a very strong tsunami. The tsunami is known from onshore deposits in Norway, on the Faroe Islands and on the eastern coast of Scotland (Dawson et al., 1988; Grauert et al., 2001; Bondevik et al., 2003). Model simulations based on hydrodynamic shallow water equations concluded with run-up heights of 3–5 m, which are roughly consistent with the tsunami wave heights deduced from geological evidence in Scotland (Harbitz, 1992). Onshore deposits in western Norway, however, indicated a run-up of 10–12 m (Bondevik et al., 1997).

In the southern side of Sweden, as well as in its eastern part facing the Baltic Sea, and its northwards continuation to the Gulf of Bothnia, a systematic paleoseismological survey revealed that a number of strong paleoearthquakes, with magnitudes up to about 8, occurred during the Holocene (see in Mörner, 2003 and a long list of relevant papers cited therein). There is evidence that some of these earthquakes were tsunamigenic (Mörner, 2005), thus making the region a natural laboratory for the investigation of paleotsunamis with geological methods.

2.3.4 Boulders and Megaclasts

At the other edge of the grain size spectrum for tsunami deposits there are boulders and megaclasts. Historically, boulder transportation onshore Santorini Island due to tsunami action was reported in documents describing the September 29, 1650 (OS) extra-caldera volcanic eruption in the submarine edifice Columbo. The study of recent events proved, in fact, that detachment of large boulders from the near shore and their transport and deposition further inland can occur due to tsunami inundations (Bourgeois and MacInnes, 2010; Paris et al., 2010, and references therein). The chronological control of the time of emplacement of boulders and megaclasts is based on dating of marine biogenic encrustations collected directly from the boulder.

Large accumulations of boulders weighing up to 80 tons at an elevation from 1.8 m to 5 m a.s.l. along the southeastern coast of the Apulian region, Ionian coast of Italy, were interpreted by Mastronuzzi and Sansò (2000, 2004) and Mastronuzzi et al. (2006, 2007) as being related to tsunami waves. By using morphological indicators, radiocarbon dating and archeological estimates those authors were able to propose the December 5, 1456 and February 2, 1743 earthquakes occurring in Apulia as the ones responsible for the generation of two large tsunamis. Similar observations were also presented from southern France (Vella et al., 2011).

Sedimentological and geomorphological evidence, credibly of tsunami origin, was documented by field observations performed in North West Crete Island by Scheffers and Scheffers (2007). They observed that in Phalasarna and Balos Bay, boulders weighing up to 50 tons at 5 m a.s.l. were interpreted that moved from the foreshore inland by tsunami waves dated around AD 365 or later. The minimum wave run-up has been estimated equal to 6 m, making the prospect of such immense movement moderate to high. More impressively, in Balos dislocated boulders of 67 tons and 75 tons at 15–25 m a.s.l. indicated a high probability of tsunami run-up of more than 25 m but further verification is needed there.

Enigmatic boulder clusters on the southern coast of Lesvos Island, North East Aegean Sea, were observed by Vacchi et al. (2012) who suggested that these clusters possibly were the result of tsunami action. However, those authors were unable to infer the origin of boulders, that is, tsunami versus storm waves, on the basis of hydrodynamic simulations. On the other hand, the suggestion that either the seismic, local tsunami of July 23, 1949 in the nearby Chios Island or the large tsunami of July 9, 1956 in Cyclades, South Aegean Sea, were responsible for the boulder accumulation is still weak given that both tsunamis were of very low height in southern Lesvos, even if it had reached there at all.

In northern Lebanon, easternmost side of the Mediterranean basin, evidence of megablocks left by extreme waves around Tripolis and Byblos were presented by Morhange et al. (2006). The older event was dated at 3639–3489 cal. yr. BP thus placing the event in the mid-Holocene. Other events were dated in historical times at 1436–1511, 1528–1673, and 1690–1950 cal. AD. None of the periods of major uplift of coastal areas coincide with the megablock dates, suggesting that the tsunami waves derived from outer tsunamigenic areas (Morhange et al., 2006). Such tsunamis, however, were produced by large earthquakes offshore Rhodes, eastern Hellenic Arc, in AD 1481, 1609, and 1741. The correlation between historical documentation and onshore geological signatures in South West Turkey was documented by Papadopoulos et al. (2005b, 2012a). Since the dates of these documented tsunamis coincide with the megablock dates in Lebanon, it would be of particular interest to examine through numerical simulation experiments whether tsunami features may account for the hydrodynamic forces required to move ashore the megablocks in the particular coastal spots. In relation to this, it is noteworthy that geomorphic evidence along coastal sections of South Cyprus and ^{14}C dating results indicated possible tsunami activity occurring between AD 1530 and 1821 but no historical correlation was found so far (Whelan and Kelletat, 2002; Noller et al., 2005). The 1609 and 1741 strong tsunamis in the eastern Hellenic Arc might provide an explanation.

Geomorphic features, including large boulders and boulder ridges, abrasion of soil and vegetation still visible in the Cabo da Roca-Cascais area, west of Lisbon (Scheffers and Kelletat, 2005), and boulder deposits at Cabo de Trafalgar, southern Spanish Atlantic coast (Whelan and Kelletat, 2002), were interpreted as possible relics of the large tsunami of November 1, 1755.

2.3.5 Geomorphological Imprints

Field observations performed after modern tsunamis, particularly the 2004 Indian Ocean tsunami and the one that hit North East Japan on March 11, 2011 have shown that strong tsunamis are capable of causing significant geomorphological changes, which may include beach erosion, disruption of sand barriers and dune systems, and also erosional escarpments and large scars (Paris et al., 2009 and references therein). In order to detect and interpret a tsunami geomorphology, we should take into account the potential interplay between sand availability, embayment type, nature of the coast, accumulation space, and landward environmental conditions (Goff et al., 2009). In the Mediterranean region, the creation of washover fans, landward oriented within lagoons or coastal lakes, has been interpreted as due to the action of tsunamis (Gianfreda et al., 2001; Luque et al., 2002). Geomorphic evidence of possible tsunami action occurring in Cyprus in the time interval from AD 1530 to AD 1821 was assumed (Whelan and Kelletat, 2002; Noller et al., 2005). Although this has not been verified historically, as noted earlier strong tsunamis generated in the easternmost Hellenic Arc may provide an explanation.

Strong soil erosion because of tsunami inundation was also reported in relation to the AD 1650 Columbo tsunami in the Santorini island complex. In fact, documentary sources reported that in Perissa, east coast of Thera, previously unknown Byzantine graves dated to the seventh–eighth century AD, situated at elevation of about 3–4 m a.s.l. and at distance of ~150 m from the present shoreline, were exposed following the erosion of surface sediments (Figure 2.23).

2.3.6 Offshore Signatures

Only recently, thanks to the huge amount of data collected after the 2004 Indian Ocean and 2011 Japan tsunamis, the international community started to investigate for tsunami signatures left offshore (see review in Papadopoulos et al., 2014a). For example, Paris et al. (2010) noted that both inflow (landward) and outflow (seaward) caused intense erosion, sediment transport and deposition, ranging in size from fine sand to boulder, up to 5 km inland and 2.5 km offshore. Moreover, the amount of sediments reworked, transported and deposited offshore by the outflow was probably larger than the volume deposited inland during the 2004 Indian Ocean tsunami (Paris et al., 2010). These recent findings show that offshore there is a higher potential for recording "anomalous" events, that is, tsunamis but also earthquake-triggered mass transport deposits, with respect to the coastal environments that experience intermittent and variable deposition/erosion as well as important human disturbance (Gràcia et al., 2010; Smedile et al., 2011).

Abyssal plain turbidite deposits were found to be an important key element to investigate tsunamigenic Holocene events of great size, such as the

LBA Thera eruption and the AD 1755 Lisbon earthquake. As regards the LBA event, near bottom 4-kHz seismic-reflection surveys showed a distinct, acoustically transparent, flat-lying layer of megaturbidites up to 30 m thick and nicknamed "homogenites," occupying the uppermost part of the sediment column in topographic lows of the Southern Ionian and Calabrian Ridges (Kastens and Cita, 1981). This stratigraphic unit is characterized by an upward thinning sequence implying deposition in a single event controlled by gravitational settling. Kastens and Cita (1981) calculated that the emplacement occurred between about 4400 years and 3100 years BP, and concluded that the homogenite was deposited from sediment transport induced by the Minoan tsunami of Thera. The thick structureless homogeneous mud was later recognized in more than 50 gravity cores independently from their setting and local characteristics (Hieke, 1984; Cita et al., 1984, 1996; Cita and Aloisi, 2000).

Moreover, Kastens and Cita (1981) calculated that the near-bottom oscillating currents were above the erosion velocity of clay-sized particles, and that the pressure pulse induced by the tsunami wave was of sufficient magnitude to cause liquefaction of the sediment draping the slope. Recent numerical simulation of the LBA Thera tsunami indicated wave amplitude in the source of 40 m or more (Novikova et al., 2011). However, homogenites were not found in the Levantine Basin to the east of longitude 26°E. Kastens and Cita (1981) supported that most of the tsunami energy was directed toward the southwestern quadrants, which contradicts their assumption that the tsunami height in the source could be calculated by inverting the elevation at which pumice was deposited by the Minoan tsunami in Jaffa, Israel; meaning in essence that they adopted efficient tsunami propagation toward the southeast as well. Sakellariou et al. (2012) suggested that a large tsunami was initiated by the Thera caldera collapse and spread toward South West, thus explaining the occurrence of homogenites. Further tsunami(s) may have been triggered when massive pyroclastic flows entered in the ocean. In that case the generated tsunami(s) spread toward all directions around the volcano.

Onshore sediment deposits of the LBA Thera tsunami found by Minoura et al. (2000) in Didim and Fethiye, South West Turkey, gave no doubt that the wave propagated toward the eastern and southeastern directions. In addition, a 40 cm-thick sedimentary deposit on the continental shelf off Caesarea Maritima, Israel, identified in four cores from 10 m to 20 m water depths, was dated, and assigned to tsunami waves produced during the LBA eruption of Thera (Goodman-Tchernov et al., 2009). Dates for the tsunami homogenites bracket both the so-called "high" and "low" chronology for the Santorini eruption. It is noteworthy that the impact area of the large AD 365 tsunami in the western segment of the Hellenic Arc is nearly identical with the spatial field of the homogenite horizon. This event certainly does not fit the homogenite from a time frame stance. However, Papadopoulos (2011) suggested that the possibility that an earlier "365-type" tsunami event, acting as a triggering agent for the homogenite horizon, could not be ruled out. Geophysical surveys and sediment cores collected from the Ionian Sea showed that the 20–25 m thick megaturbidite known in the literature as Homogenite/Augias was triggered not by the Santorini caldera collapse but by the 365 AD Cretan mega earthquake/tsunami (Polonia et al., 2013).

In the western part of the Marmara Sea, a record of turbidites was obtained in five cores, and was correlated to strong earthquake shaking, in particular with the large (M_w 7.4) Ganos earthquake of August 9, 1912 which indeed triggered a local tsunami as reported by Altinok et al. (2003).

In the Augusta Harbor, East Sicily, Smedile et al. (2011) worked on a 6.7 m long, fine sediment core sampled 2.3 km offshore at a water depth of 72 m. A multivariate analysis on benthic foraminiferal assemblage showed the presence of 12 anomalous layers, marked by a high concentration of displaced epiphytic foraminifera. These layers are also characterized by subtle grain size changes and peculiar bimodal particle-size distribution. Smedile et al. (2011) suggested that these layers could be related to high-energy exceptional events able to disperse an extra amount of infralittoral epiphytic species toward deeper areas, with tsunami back-wash as the best candidate. The tsunami mechanism is supported further by the coincidence between five events ages and historical tsunamis, both local and basin-wide that hit the area. Namely, the local tsunamis of AD 1908, AD 1693, AD 1169, and the basin-wide ones of AD 365 Crete and the LBA of Santorini were proposed.

On the other side, the "turbidite paleoseismology" concept has been applied, for example in the South West Iberian Margin (Gràcia et al., 2010) and in the Calabrian Arc (Polonia et al., 2012). Sediment cores collected in the Tagus Abyssal Plain, Infante Don Henrique Basin, Horseshoe, and Seine Abyssal Plains (Figure 2.25) revealed that deep-sea basins preserve a record of episodic deposition of turbidites. In the South West Iberian Margin, excluding specific climatic events, earthquakes, and tsunamis are the most likely triggering mechanism for synchronous, widely spaced distributed turbidites during the Holocene. Age correlation together with textural,

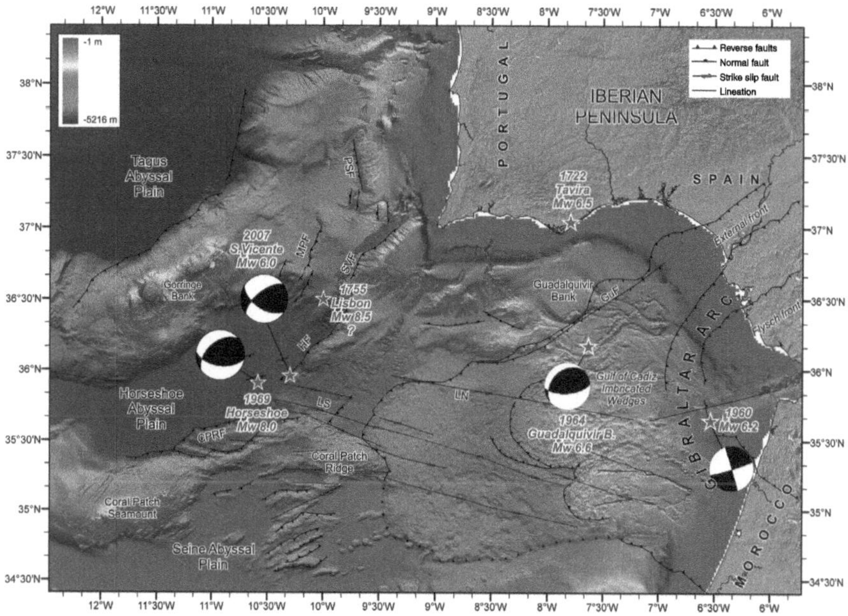

Figure 2.25 ***Color shaded relief map of the South West Iberian margin based on SRTM-3 for land topography and SWIM bathymetric compilation for bathymetry (Zitellini et al., 2009).*** Red stars illustrate epicenters of historical and instrumental strong ($M_w \geq 6.0$) earthquakes which occurred in the South West Iberian Margin, and the white/dark balls represent the fault plane solutions of $M_w \geq 6.0$ instrumental earthquakes (Fukao, 1973; Buforn et al., 1995, 2004; Solares and Arroyo, 2004; Stich et al., 2005, 2007; Baptista and Miranda, 2009). Black lines show active faults. *(Modified from Zitellini et al. (2004, 2009); Gràcia et al. (2003a, 2003b, 2010); Terrinha et al. (2009); Bartolome et al. (2012); Martínez-Loriente et al. (2013); after Papadopoulos et al. (2014a).)*

physical properties, and geochemical signature of turbidite deposits reveal a total of seven Holocene widespread turbidite events. Precise dating of the most recent turbidite event based on ^{210}Pb and ^{137}Cs geochronology provides an age of AD 1971 ± 3 (Garcia-Orellana et al., 2006), which corresponds to the high-magnitude (M_w ~8.0) AD 1969 Horseshoe tsunamigenic earthquake. Calibrated ^{14}C ages of subsequent widespread turbidite events correlate with the November 1, 1755 earthquake while paleotsunami deposits in the Gulf of Cádiz dated around 218 BC. Taking into account older synchronous events, occurring from 4960–5510 yrs BP to 8715–9015 yrs BP, a recurrence interval of about 1800 years is obtained for large magnitude earthquakes/tsunamis during the Holocene period (Gràcia et al., 2010).

In the western Mediterranean, most failures have limited volume, short runout, and originate in relatively deep water. Therefore, only the largest albeit infrequent events are likely to trigger large tsunamis. The so-called BIG'95, with a volume of 26 km^3 and 110 km runout (Lastras et al., 2002; Urgeles et al., 2007), is one of such events in the western Mediterranean basin (Løvholt et al., 2009; Tinti et al., 2009; Iglesias et al., 2012). Seismicity is a major control in the distribution, magnitude and typology of submarine landslides (Camerlenghi et al., 2010; Urgeles and Camerlenghi, 2013), but other factors, mainly fluvial sediment input and margin progradation (Field and Gardner, 1990; Lastras et al., 2007), come to play an important role too.

2.4 INSTRUMENTAL RECORDS

Instrumental records of tsunamis in the European–Mediterranean region are available only for a very few events. This is due to the low rate of tsunami occurrence, which means that only very few tsunamis have occurred since the first installation and operation of tide gauge by the end of the nineteenth century. At all evidence, the first tsunami recorded by tide gauge was the one caused by the strong earthquake of February 23, 1887 (Eva and Rabinovich, 1997; Larroque et al., 2012). Three tide gauge records are available for the large tsunami of July 9, 1956 produced by a magnitude 7.4 tectonic earthquake in the South Aegean Sea. Tide gauge records are also available for some local tsunamis such as those caused by the large *M* 6.8 earthquake of February 24, 1981 in the eastern Corinth Gulf, Central Greece, and by a volcanic landslide during an eruptive episode in the volcano of Stromboli, Aeolian Arc, on December 30, 2002. The earlier mentioned tide gauge were utilized by several authors in their effort to understand the 1956 tsunami source mechanism as analyzed in Chapter 4. The May 21, 2003 small-to-moderate Boumerdes–Zemmouri (Algeria) tsunami produced was also recorded by tide gauge instruments. Of great importance for understanding source mechanisms are also the seismic records of the earthquakes that caused tsunamis in about the last century or so. The May 21, 2003 small-to-moderate Boumerdes–Zemmouri (Algeria) tsunami produced was also recorded by tide gauge instruments

Collections of photographs illustrating the impact of tsunamis are valuable sources of information. Such collections exist for large tsunamis, such as the Messina of December 28, 2008 (Società Fotografica Italiana, 2002; Bertolaso et al., 2008) and the Amorgos of July 9, 1956 (Galanopoulos, 1957a,b).

CHAPTER 3

Impact of Tsunamis

3.1 INTRODUCTION

The systematic study of the impact of tsunamis in the European–Mediterranean region has been rather neglected so far. This is due to the low rate of tsunami occurrence in the region, which implies that in the last century or so only very few important tsunami events took place. However, in the historical period there are a number of tsunamis for which the available documentary sources offer adequate information to study the many different effects of the tsunami events.

In this chapter an effort has been made to systematically study the tsunami impact in the European–Mediterranean region from the ancient times up to the present. Prehistorical tsunamis were excluded since no written evidence is available. For example, the large tsunami produced by the big volcanic eruption of Thera in the South Aegean Sea by the end of the seventeenth century BC constantly attracts great interest since the eruption and its associated events, such as the tsunami, at all evidence influenced directly the progress of the LBA civilizations in the East Mediterranean. However, the possible impact of the so-called Minoan tsunami only indirectly is inferred since very little direct archeological evidence is available.

A number of selected tsunamis occurring in the European–Mediterranean region were used with the aim to better understand how the tsunamis affected both human society and the natural environment. The selected tsunamis were chosen with the following criteria: (a) their source areas cover the entire region under examination; (b) the descriptions available contain comprehensive information as regards the tsunami impact; (c) not only large but also local tsunami events were selected; and (d) a variety of tsunami source types were included, that is seismic, volcanic, and landslide sources.

The number of historical documentary sources and other publications that were used and reviewed to collect the relevant information is very large and, therefore, it is impossible to list all of them. However, one may refer to these scientific publications which offer more detailed citations: Galanopoulos (1960), Ambraseys (1962, 2009), Jørstad (1968), Antonopoulos (1979, 1980), Tinti et al. (1988), Papadopoulos and Chalkis (1984), Papazachos et al. (1985), Soloviev (1990), Guidoboni et al. (1994), Tinti and

Tsunamis in the European-Mediterranean Region. http://dx.doi.org/10.1016/B978-0-12-420224-5.00003-X

Maramai (1996), Altinok and Ersoy (2000), Soloviev et al. (2000), Guidoboni and Comastri (2005), Salamon et al. (2007, 2009), Papadopoulos (2003b, 2009, 2011), Yalciner et al. (2002), Tinti et al. (2004), Fonseca (2005), Papadopoulos and Fokaefs (2005), Fokaefs and Papadopoulos (2007), Papadopoulos et al. (2007a, 2010, 2011), Baptista and Miranda (2009), Altinok et al. (2011), Maramai et al. (2014), Papadopoulos et al. (2015), and Diakogianni et al. (2015).

3.2 EFFECTS OF SOME IMPORTANT TSUNAMIS

3.2.1 426 BC, North Evoikos Gulf, Central Greece

This was a rather local tsunami caused by a strong earthquake in the tectonic rift of North Evoikos Gulf (Figure 2.8). At the time of the earthquake the sea at Orobiae (Rovies) receded from the shoreline, then a great sea wave engulfed part of the city. Many people perished. A similar flood occurred in the island of Atalanti carrying away part of the Athenian fort and wrecked one of two ships, which had been drawn up on the shore. In Peparethus, today Skopelos Island, North West Aegean Sea, the sea was reported to recede but there was no flood.

3.2.2 July 21, 365, Western Hellenic Arc

This was a basin-wide tsunami produced by a very large earthquake that ruptured at the western Hellenic Arc (Figure 2.2). In Alexandria, huge masses of water killed thousands of people (some sources say 5000 people). Ships in the harbor were raised to great heights, overtopping high buildings and walls and removed inland among the yards and the houses. Some ships were destroyed by the rapid whirlpools created by the retreating waters. Some great ships were hurled by the fury of the waves onto the rooftops of Alexandria. Others were thrown up to 2 miles from the shore. The wave inundated the area of Panephysis, near the modern El Manzala Lagoon to the east of Alexandria in the Nile river mouth, where it seems it penetrated a long distance inland and turned fruitful lands into salt marshes. Boats on the Nile were drifted ashore by the water of the river up to 18 stades (ca. 3.5 km) inland. The sea flowed over the shore, causing suffering to countless people in Sicily and many other islands. In Areopolis (Palestine), the sea swept in over the shores there and in many other islands. In Epidavros, modern Cavtat near Dubrovnik in Dalmatia, Adriatic Sea, ships were hurled onto the rugged mountains and remained suspended there.

In the Adriatic Sea, also in Alexandria and in the Aegean Sea, the deep seabed was laid bare, and many sea creatures could be seen embedded in the mud. Great ranges of mountains and valleys gazed upon the rays of the sun. Many ships were stranded as if on dry land. Growling waters rose up in their turn and burst through the seething shallows flattening many buildings in the towns. Sediment layers were deposited in Phalasarna, North West Crete (see Chapter 2).

3.2.3 544 or 545, Bulgarian Black Sea

A large earthquake occurring offshore of the Bulgarian coast of the Black Sea (Figure 2.2), very likely produced a strong tsunami that inundated in three coastal zones, namely Odessus, later Tiveriopolis and today Varna; Dionysopolis, today Balchik situated 20 km to the north of Varna; Aphrodisium, situated in the Thracian Peninsula at the Xeros (today Saros) Bay, Northeast Aegean Sea. It is reported that the sea advanced on Thrace by 4 miles [ca. 6 km] and covered the three territories. Many were drowned in the waters. However, no further details are available. Besides, it is puzzling how a single tsunami event was capable to inundate coastal zones in the Bulgarian Black Sea and at the same time Thracian coasts facing the North Aegean Sea. This tsunami event is a good example, which indicates the several uncertainties that often arise from the historical documentation of tsunamis.

3.2.4 556, August, Kos Island, Dodecenese Island Complex, Greece

There was a large destructive earthquake that hit the island of Kos (northeast side of segment 10 in Figure 2.1). It is reported that the sea rose up at great height and inundated the buildings along the coast causing destruction of human beings and to their property. The size of the inundation was so big that everything that the sea could not ride over was pulled down and destroyed. Nearly all the inhabitants perished without any discrimination, including those who got refuge in temples, in houses, or in other settlements. Also, the report says that, in addition to other disasters, the entire local water supply became salty and undrinkable because it was contaminated with seawater. It is not clear if this happened due to the strong earth shaking, to the inundation of the coastal zone or both.

3.2.5 January 18, 746, Levantine Sea

A very large earthquake ruptured at the region of Palestine (segment 18 in Figure 2.1) and caused destruction and several ground failures in a large area. One of the documentary sources says that an extreme storm occurred in the sea, with the waves rising up to the sky. The sea boiled like a cauldron in the fire and the waves surged with a horrible and terrifying noise. The boiling sea overflowed and destroyed most of the cities and villages along the coast.

3.2.6 May 20, 1202, Cyprus and Levantine Sea

A large number of documentary sources indicate that this was a very large, highly lethal earthquake, particularly in Syria, Lebanon, and Israel. The earthquake very likely ruptured in the Middle East along the Dead Sea Fault Transform (segment 18 in Figure 2.1) and was felt across an area of at least 500 km from Cairo to Lesser Armenia and from Cyprus to eastern Syria. Gigantic waves rose up in the sea between Cyprus and the coast of Syria. The sea withdrew from the coast, boats smashed on to the eastern coast of Cyprus, fish were thrown on to the shore, and lighthouses were damaged.

A puzzling point is how a tsunami could be triggered from an earthquake that at all evidence was not generated by submarine earthquake but by an inland one. This issue is examined in the next chapter.

3.2.7 August 8, 1303, Eastern Hellenic Arc

This was another basin-wide tsunami caused by a large earthquake that ruptured at the eastern Hellenic Arc (Figure 2.2). In Heraklion, it destroyed buildings and killed inhabitants. In Alexandria and Bab-Al-Bahr, Egypt, the sea inundated the coastal zone and caused destruction to the port facilities. In the Nile, boats were shifted inland. The wave reached to Acre, Palestine/Israel, where the sea flooded the shore while people were swept away and drowned by the huge wave. Some authors reported tsunami inundation in Rhodes Island but this was not verified from the available documentary sources. Three layers of tsunami sediment deposits were found in Dalaman (South West Turkey) (Papadopoulos et al., 2005b, 2012a). One of them was attributed to the 1303 tsunami.

3.2.8 October 14, 1344 (or 1343), Marmara Sea

A large earthquake that very likely ruptured the north side of Marmara Sea (Figure 2.13), triggered a strong tsunami. In some places the wave swept away a number of vessels that were in the harbor or sailing near the coast, and crashed them on land. Many people perished. When the sea retreated, a lot of fish were left on the beach. The large sea wave caused great destructions in Istanbul and in several other cities of Thrace in Marmara Sea. The sea receded, leaving mud and dead fish on land.

3.2.9 September 10, 1509, Marmara Sea

A large earthquake shook the Marmara Sea region (segment MS in Figure 2.1). At Constantinople, during the shock the sea rushed inland, over the walls, and flooded the streets in the city as well as the settlements outside the walls. The district of Galata was heavily damaged by the sea wave that caused many casualties. Near Galata and near Hereke Bay (north of Izmit Gulf) the coasts were flooded. In Yenikapi, the waves overtopped the sea walls whose crown height is about 6 m. Aksaray was flooded. The Izmit shipyard collapsed and waves flooded the dockyard.

3.2.10 March 9, 1630, Western Hellenic Arc

A large earthquake that very likely occurred in the western Cretan Sea (segment 9 in Figure 2.1) caused extensive destruction and some victims in Heraklion, Crete Island. Based on a series of documents from Zakynthos Island archival sources, it is deduced that a strong tsunami was produced. The tsunami was observed around the straits of Kythira Island. The available descriptions leave no doubt that the ships sailing in that area were in peril due to strong tsunami waves traveling South and South East. According to eyewitness accounts, remnants of shipwrecked vessels and persons were observed in the sea. At the harbor of Kythira, today Kapsali, located on the southern side of the island, a slight inundation was observed at the pier.

3.2.11 September 29, 1650 (Old Style), Kolumbo Volcano, South Aegean Sea

It was reported that after the paroxysmal phase of the submarine eruption of Kolumbo (Figure 2.3), a sea swell encircled the whole of Thera Island and that the tsunami was characterized by a withdrawal and then inundation of the eastern coast to a distance of a probably exaggerated figure of 2 miles. The tsunami is reported to have swept away churches, enclosures, boats, trees, and agricultural land (Figure 3.1). Large submarine boulders are reported to have been transported from the near-shore zone and deposited inland. The coastal plains and valleys were littered with marine pebbles and dead fish. In Perissa, previously unknown Byzantine graves of the eighth century AD were exposed following the erosion of surface sediments (Figure 3.2). Tsunami sediment deposits were left behind in the archeological site of Perissa (see Chapter 2). Contemporary accounts state that the intercaldera coast was not significantly affected. It was reported that on the East and West coast of Patmos Island, the sea rose 164 ft. (50 m) and 98 ft. (30 m); on Ios Island, the waves rose to 50 ft. (15 m); while at

Figure 3.1 ***The Byzantine church of St. Irini at the coast of Perissa. It was reported that the church was hit by the AD 1650 tsunami.*** *(Photo courtesy by G.A. Papadopoulos.)*

Sikinos Island, the tsunami swept 350 ft. inland. A number of fishing boats moored at Heraklion, north Crete, are also reported to have been sunk by the tsunami when the wave overtopped the city harbor walls being of a height of about 4 m.

3.2.12 January 11, 1693, Eastern Sicily

After a very strong earthquake (Figure 2.15) at Catania, a remarkable sea level rise and withdrawal that carried a lot of boats occurred. The seawater inundated the town penetrating into the San Filippo square; farms located near the town were inundated. At Augusta, the first sea recession drained the harbor completely, causing severe damage to the ships. During the following inundation, the sea submerged the district close to the port as far as the S. Domenico monastery and killed a number of people. At La Cittadella of Messina, lots of fish were found on the beach. At Agnone, the sea withdrew and then returned violently, exceeding its usual limit. At Mascali, the sea flooded the shore for about 1 mile inland. At Taormina, the sea withdrew for about half a mile dragging some small boats. At Gozo, Malta Island, the sea at Xlendi rolled out to about 1 mile and swept back a little later.

Figure 3.2 ***View of the early Byzantine archeological site of Perissa that was revealed due to the soil erosion caused by the AD 1650 tsunami.*** Tsunami sediment deposits were found there (see Chapter 2). *(Photo courtesy by G.A. Papadopoulos.)*

3.2.13 January 7, 1731, Storfjorden, Norway

On January 7, 1731, a large rockslide occurred on the mountain Rammerfjell in Stranda, South West Norway. It was estimated that more than 100,000 m^2 of rocks fell into the fjord Storfjorden. The tsunami that was produced caused considerable damage at Stranda on the other side of the fjord, about 2 km from the slide area. This is one of the most important tsunamis historically reported in Norway given that the run-up height above the high-water level exceeded 100 paces, that is 70 m. Seventeen casualties were reported.

3.2.14 November 1, 1755, South West Iberia

This is one of the largest tsunamigenic earthquakes in the European-Mediterranean region in the historical period (Figure 2.18). The earthquake caused extensive destruction in Lisbon and other places in Portugal and beyond. The tsunami that was produced flooded Lisbon where it caused great damage and many victims. Along the south Portuguese coast (Algarve) several dwellings disappeared. In Albufeira, which stands on a relatively high cliff, people took refuge along the beach, which they thought would be safe. However, the sea attacked and swept them all away. In the town of Portimão, the force of the waves was frightening. At Lagos, the sea flooded the shore destroying the city walls. Buildings disappeared without trace. In Mazagan (El Jadida), Morocco, the sea entered the open ground of the fortress, where it left many fish when it retreated. The total number of casualties due to the tsunami may have exceeded 900. Tsunami sediment deposits were left behind mainly in the coastal zone of Algarve (see Chapter 2).

3.2.15 February 6, 1783, Messina Straits

Another very strong earthquake ruptured the area of Messina Straits (Figure 2.15) and generated a very powerful tsunami. At Scilla, waves reached the roof of the buildings and more than 1500 victims were noted. The sea inundated the district of Marina Grande, reaching the first floor of buildings, then in the cliffs of the Castle at Monasina. The water climbed the Livorno stream up to 200 m from the seashore, then reached Chianalea and Marina dell' Oliveto, located on the other side of the promontory. Three big inundations were observed. At Marina Grande, the Santo Spirito church was destroyed, the church of Santa Maria delle Grazie, 90 m from the shore, was inundated as well as the San Nicola church. A woman became entangled in a mulberry tree at a height of 8.3 m near the Gornelle fountain. At Nicotera, the sea withdrew then inundated the shore carrying some fishing boats. The inundation was also noted in Bagnara and Cannitello. At Messina, the sea was seen to rise about 2 m and overflow its coast with noise. At Peloro, near Torre del Faro, flooding of a long stretch of fields was observed, close to the Pantanello small lake. Small houses, people, and animals were carried seaward. At Torre del Faro, the coast flooded for about 600 paces. Some boats and 26 victims were carried seaward. At Punta del Pezzo and surrounding areas, the sea covered the beach for one and a half miles, leaving some sand on the ground. At Reggio Calabria, the sea inundated the shore carrying silk equipment.

3.2.16 February 23, 1887, Ligurian Sea

A series of strong earthquakes shook the area of Ligurian Sea (segment 2 in Figure 2.1). During the third shock, the harbor nearly dried, then the sea returned violently. At Sanremo, some vessels broke moorings, while lots of fish were found on the beach. At Savona, the sea withdrew and lots of fish were found on the beach. At Antibes and Nice, the sea level lowered and the sea bottom became dry. At Loano and between Imperia and

Ospedaletti, permanent coastal changes and associated shoreline regression were observed.

3.2.17 December 28, 1908, Messina Straits

A very strong earthquake (Figure 2.15) caused great destruction and nearly 60,000 victims in the Messina Straits. Severe damage to harbor infrastructures, boats, and houses was also caused by the tsunami that followed the earthquake. About 1500 victims were reported. All piers in the harbor of Messina lowered and some of them collapsed. Most of them were lying in the southern beach of the city, close to the Portalegna stream. The English cemetery was completely destroyed. Many boats were carried onshore and destroyed. Many vessels moored in the harbor felt a sudden considerable sea level rise just a few minutes after the mainshock, and some of them broke their mooring. At least three huge waves were observed, the first being the biggest.

The tsunami involved mainly the eastern Sicilian coast but also the southern and northern ones. At Paradiso, north of Messina, many boats were carried to the beach. Many small houses were ruined while three victims were reported. At Galati Marina, many buildings were destroyed. A lot of big trees were uprooted. Many ships were carried onshore and damaged and 14 people were killed. At Briga Marina, many houses were destroyed and 46 people were drowned. In the San Paolo district, the damage was heavier than in other places in Sicily. At Giampileri Marina, many walls collapsed. At Nizza di Sicilia, many walls and boats were destroyed. At Alì Marina, many boats were damaged and trees were uprooted. At Giardini Naxosmax two people were killed and many boats were damaged. At Riposto, the sea flooded the coast about 150 m, 13 people died and many walls collapsed. At Aci Trezza, three waves were reported, many trees were uprooted, and boats were destroyed. At Catania, three waves were observed, many boats were damaged and in the wall of the tide gauge station it was possible to distinguish four different wave marks, and three children were killed. At Augusta, the water flooded more than 700 m of land and severe damage was caused to the salt pan. At Cefalù, very light inundation, of less than 10 m, happened. The Calabrian coasts facing the Messina Straits were severely affected by the tsunami. In particular, at Reggio Calabria many houses were razed to the ground, harbor infrastructures were severely damaged, many boats and vessels in the harbor were damaged, and the railways were completely destroyed. At Gallico, a very large inundation of about 400 m occurred. At Pellaro, most of the houses close to the beach were totally destroyed, and many people were carried landward. At Lazzaro, the village was almost completely destroyed, and 90 people were killed, while beach erosion happened. In many localities such as Scilla, Nicotera, Tropea, Scalea, Bianco, Bova Marina, Porto Salvo, the sea flooded the beach for many meters. In the Aeolian Islands, the tsunami was only weakly felt. In Malta Island at Sliema and Misida the water inundated the streets and some shops in the square. At Marsalforno many submarine cables were broken.

3.2.18 February 9, 1948, Karpathos Island, Eastern Hellenic Arc

A very strong earthquake caused destruction in Karpathos Island (segment 10 in Figure 2.1). In Pigadia Bay, the main port of Karpathos, the South East side of the island, the writer Mrs Fotini Makri (born in 1915) as well as other eyewitnesses reported to the author that the first tsunami wave arrived about 5–10 min after the earthquake. The wave overtopped the fountain, penetrated inland up to Mrs Makri's house, which is about 50 m from the fountain or about 60 m from the shoreline. Seaweed was left behind. Eyewitnesses' accounts, as well as pictures of Pigadia Bay published by Mrs Makri, indicate that the fountain was elevated on a wall about 2.5 m above the mean sea level. Many vessels were moved ashore and destroyed. Houses along the Pigadia coast were flooded and doors were broken by the wave. The sea disturbance continued for several hours. Further to northwest, around the oil press of Mr M. Sarris and in the mouth of a nearby small local river, the ground is flat and slightly elevated above the sea level, which makes the area prone to tsunami inundation. In fact, eyewitnesses reported that the 1948 wave penetrated inland about 250 m, while fish were left behind up to a distance of about 200 m from the shoreline.

3.2.19 July 9, 1956, Cyclades Islands, South Aegean Sea

This was the largest crustal earthquake that occurred in Greece (Figure 2.3) in the last century or so. Some minutes after the earthquake, a large tsunami inundated many coastal zones in the islands of the South Aegean Sea. More than 80 small ships and row-boats were wrecked, sunk, or vanished. Four persons were killed and a few others were injured. In the island of Amorgos at Katapola, in the north side of the island, the first sea movement was a flooding and the wave was about 3 m high, causing severe damage in the harbor. Many houses close to the shoreline were heavily affected. Four large

Figure 3.3 ***A 5-ton fishing-boat is attacked by the tsunami in Porto-Scala of Astypalea Island.*** See also in Figure 3.4. *(Photo courtesy by anonymous.)*

vessels were thrown onto the quays and four others were wrecked on shore. The beach near the town and the neighboring fields were littered with trunks and limbs of trees; many fields were washed up.

In the island of Astypalaea in many localities hundreds of acres of cultivated land was washed away. At Astypalaea, an initial withdrawal was followed by a flooding (inundation 400 m) with a maximum height of about 20 m. At Porto-Scala, a large number of fishing boats were crashed on shore and debris was carried over 500 yards inland (Figures 3.3 and 3.4). Many houses were damaged and two people were injured. In Kalymnos Island, the tsunami caused the greatest damage because the main settlements of the island were developed along the shore, directly facing the wave front. In the town of Kalymnos, the wave was felt more strongly, with severe damage, with the first movement of withdrawal being of about 5 m. The maximum height was about 3.5 m. In the town of Pothea, the sea first withdrew for about 2.5 m and then rose flooding the entire town. The wave impulse was so severe that blocks of residential buildings in the harbor as well as a small ship building installation 100 m inland were damaged. More than 30 fishing boats and one large vessel had drifted ashore and crashed (Figure 3.5). The sea receded and advanced three times in the first hour, carrying debris and

Figure 3.4 ***A general view of Porto-Scala in Astypalea. Eyewitnesses reported to the author that the fishing boat shown in Figure 3.3 was thrown up inland at an elevation of about 11 m at the point indicated by an arrow.*** *(Photo courtesy by anonymous.)*

Figure 3.5 ***Large fishing boat moved ashore in Kalymnos Island from the July 9, 1956 tsunami (Galanopoulos, 1957b).***

furniture 1 mile inland. Three persons drowned. At Kantouni, extensive areas of cultivated land at different places of the island were washed away.

In the island of Antiparos, the sea overflowed the shore and also the town of Kastro, which stands on a ridge near the northern end of the island. Buildings near the coast suffered little damage, the quay was washed

Figure 3.6 ***The coastal zone of Lakki harbor in Leros Island covered entirely by water during the inundation of the July 9, 1956 tsunami (Galanopoulos, 1957b).***

away and other parts of the dock completely settled. A motorboat was taken by the waves and crashed on the rocks 15 m inland. On the island of Lipsos many houses were damaged and one collapsed. In the harbor the wharf and the quay were demolished and harbor facilities were severely damaged. In Leros Island, several fishing villages were flooded (Figure 3.6) while many houses around the harbor of Lakki village were damaged and goods stored on the wharfs were taken away by the waves. Many fishing boats were destroyed. At Finiki, in the South West side of the island of Karpathos, the wave was so severe that it demolished the breakwater and anchorage of the harbor. In the island, many villages were flooded, with severe damage to houses. Thirty cattle were carried away to the sea. The damage was much less in Crete, Patmos, and Ikaria islands although the wave inundated there too (Figure 3.7).

3.2.20 February 7, 1963, Western Corinth Gulf

After a coastal landslide without earthquake (Figure 3.8), a strong local tsunami attacked the south and north coasts of the western Corinth Gulf, Central Greece. Two persons were killed, and another 12 were injured. Destruction was reported in houses, cultivated land, fishing boats, and other property. Fish were found after the retreat of the sea.

Figure 3.7 ***In Patmos Island small vessels stranded on dry land after the seawater withdrew during the July 9, 1956 tsunami (Galanopoulos, 1957b).***

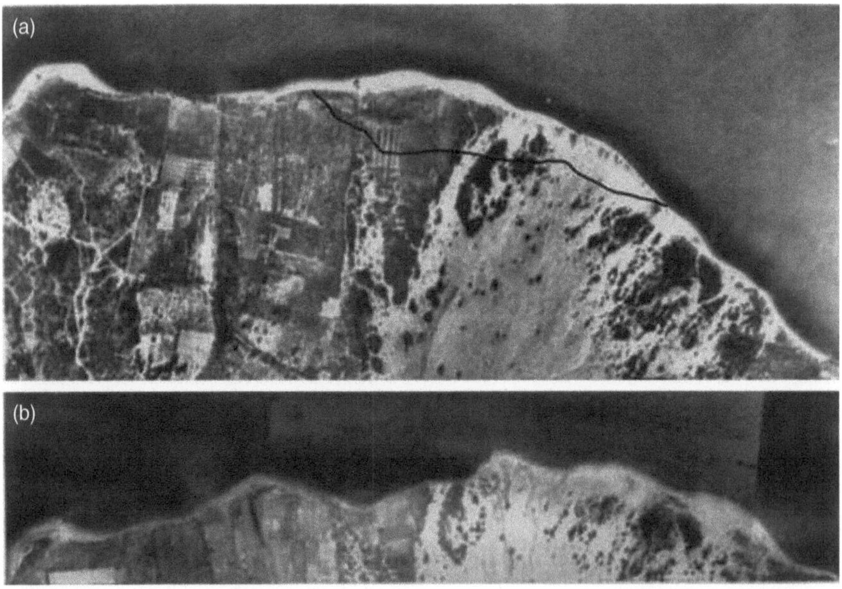

Figure 3.8 (a) Air photographs taken before and (b) after the landslide-generated tsunami in west Corinth Gulf on February 7, 1963 (Galanopoulos et al., 1964). Black line shows the landslide sediment mass.

3.2.21 August 17, 1999, Izmit Bay, Marmara Sea

After a very large earthquake that ruptured at the western segment of the North Anatolia fault (Figure 2.13), a local tsunami was observed in the Izmit Bay, on the eastern side of Marmara Sea. In almost all places the tsunami was seen to cause the sea to withdraw first and then to overcome the usual shoreline and inundate land. In some places boats and vessels were carried on land. The tsunami was exceptionally violent in the town of Degirmendere. Here, the sea receded, the wharf collapsed, vessels were drifted with the receding sea, thrown back about 150 m, and lifted up with a proceeding great wave. Two piers, a hotel, a restaurant, a cafe, and 14 trees at Cinarlik Park were swept away to the sea. The wave dragged seven people near the seaside and only two of them survived. Buildings and restaurants along the coast sank into the sea along with cars. At Aksa Factory, a platform 60 m^2 and 75 m away from the coast, completely vanished. Moss and dead fish were found along 70–100 m coastal band of the Laledere delta, showing inundation.

At Dilovasi Port, the first sea movement was a withdrawal, while some ships were uplifted. At Sirinyah, the first floors within 5 m to the sea were flooded. Some seabed material were carried by the wave into the houses. At Cene Suyu (Izmit), a fisherman sitting by the coast claimed that the sea receded by about 30 m following the earthquake and then a giant wave rushed in, causing the total destruction of the jetty. The man was pulled into the sea. Then the sea receded again and he was thrown back by a second wave. In Izmit Marina, the moored boats fell down 2 m.

3.2.22 December 30, 2002, Stromboli

Following a volcanic landslide during an eruptive episode in the volcano of Stromboli, a destructive tsunami, starting with an initial slow withdrawal, violently attacked the northern coasts of the island with maximum water height of about 9 m (Figure 3.9). Three or four strong waves were reported. It has been ascertained that within a few minutes the water invaded Ficogrande Beach, and attacked Punta Lena and Scari producing severe damage. The strongest tsunami effects were observed at Stromboli. At Piscità, the buildings located near the shore were seriously damaged. Typical effects of the tsunami impact were low brick walls and balustrades pulled down, door and window frames unhinged, windowpanes broken, sand deposited inside the houses and along the inner narrow streets, shrubs flattened, small boats and other objects moved inshore. At Ficogrande, the damage produced by the tsunami was very severe. Some small buildings

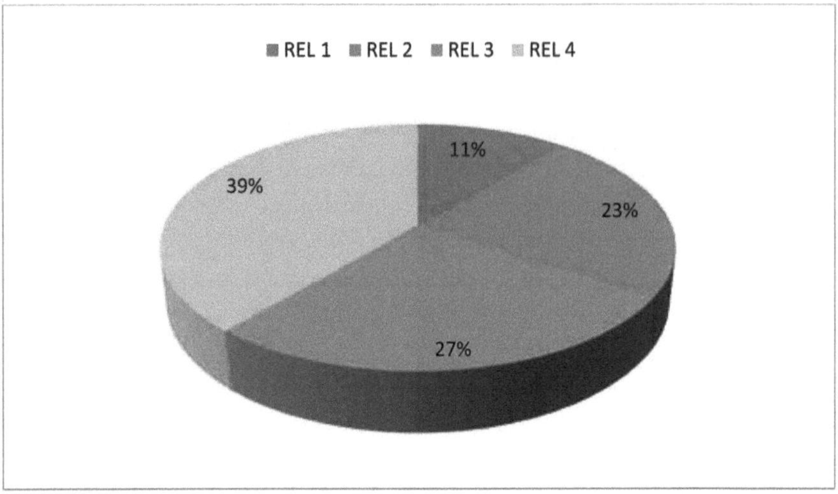

Figure 3.9 ***Distribution of reliability assignment of the tsunami events listed in the new tsunami catalog for the European–Mediterranean region presented by Papadopoulos et al. (2015), Diakogianni et al. (2015).***

were completely destroyed and a lot of walls were knocked down. A large amount of sand was deposited in courtyards, terraces, and inside the houses. Heavy objects (i.e., lava blocks, boats, scooters, etc.) were transported by the wave.

At Punta (or Sopra) Lena, party walls, gates, doors, and windows were pulled down or destroyed. Railings and iron fences were bent. Pebbles thrown shoreward by the waves remained stuck inside the slots of several windows. Small pebbles were also found over the roof of a building at the height of about 7 m. Heavy objects, such as gas cylinders, household appliances etc., were dragged by the wave. Furniture inside several houses was untidily heaped or even ejected outside due to strong whirlpools. Scari is the place where the maximum horizontal inundation (146 m) was measured. However, the tsunami impact was not important due to the wide beach as well as to the relatively large distance from the tsunami source. Some damage was suffered by buildings, such as warehouses, boats depots, and small huts in the northern side of the beach. Particularly impressive was the erosion caused by the tsunami on the house foundations and the possible formation of erosional scarps. At La Petrazza, the tsunami was responsible for erosive phenomena associated with small landslides of the steep near-shore escarpments. At Ginostra, eyewitnesses reported strong sea agitation in the little harbor where two small boats were carried onto dry land. At San Pietro harbor, damage was reported to the facilities on the beaches (e.g., kiosks, tables, etc.), while small boats were carried in land.

3.2.23 May 21, 2003, North Algeria

A strong, damaging earthquake ruptured at the north margin of Algeria. Because of the tsunami that was created, many hundreds of vessels in various harbors were damaged. Fishermen reported that in the port of Zemmouri Bahar the water depth in the port dropped to a height of less than 1 m and that several fishing boats ended up resting on the bottom of the port seabed before the water came back to its original position. In the Balearic Islands, about 180 boats and vessels in the various harbors were damaged or sunk. At Majorca Island in Palma, the first sea movement was an ingression and the Paseo Maritimo street was flooded. More than 100 boats were destroyed. At Cala Agulla, the sea flooded the beach. At Porto Cristo Manacor, the 'Riuet area' was flooded, restaurants were flooded, and boats were affected. At Cala d'Or, the road was flooded and the pier was destroyed, while the Porto Petro was flooded and some boats sunk. At Santanyi, one boat sank, at Andratx the harbor was damaged, and at Ses Salines six boats sank. At Cala Galdana, the harbor was damaged, and at Sant Lluis the boats sank. At Cala Llonga, boats sank and the pier was destroyed. It was reported that at Mahon harbor 11 boats sunk, 110 boats were damaged, and a pier was destroyed. At Sant Antoni de Portmany, a pier was destroyed and some boats sunk, while in Dr Fleming Avenue, 15 cars flooded. At Santa Eulalia, the pier was flooded, two boats sank and a wave height of 1.5 m was observed. At Badia de Portmany, the sea withdrew for about 100 m, then it flooded the shore causing damage to boats and cars.

In Formentera Island at Es Pujols some boats sank. In the Figueirette harbor, the water level lowered about 1.50 m causing damage to some boats.

3.3 IMPACT STATISTICS FROM A NEW TSUNAMI CATALOG

Although the impact of the tsunami events, which are known in the European–Mediterranean region, has been described in a large number of documentary sources, no systematic study has been made so far aiming to understand what are the main types of impact, the level of impact, as well as their space and time distributions. Very recently, our team at the National Observatory of Athens performed such a study the main results of which are presented here. The first step was to further improve the tsunami catalog covering the entire European and Mediterranean region from prehistoric times up to the present. The catalog is of increased completeness and homogeneity with respect to previous ones, containing more than 520 events with reliability assignment for 370 of the events listed (Figure 3.9). Following the tradition established with the organization of the New European Tsunami Catalogue since the 1990s with the EU-supported research projects GITEC (1992–1994), GITEC-TWO (1996–1998), and TRANSFER (2006–2009), the reliability of each event has been chosen to scale from 1 (lowest reliability) to 4 (highest reliability). New historical events were inserted in the catalog, while revised parameters of historical tsunamigenic earthquakes were extensively adopted particularly for the most active region of the eastern Mediterranean.

The distribution of the tsunami sources in the entire European and Mediterranean region is illustrated in Figure 3.10 where the maximum intensity assigned to each tsunami event is shown. Intensity integrates into one figure the several types of impacts produced by a single tsunami event and is a proxy of the tsunami size. As one may observe the majority of the tsunami events as well as the largest of them are concentrated in the eastern Mediterranean basin and particularly along the Hellenic Arc.

In association with the catalog, an inventory was created containing descriptions of the impact that each tsunami event had. Some examples were given in the previous section of this chapter. In order to secure the high reliability of the information used, we considered only tsunami events assigned with reliability 3 or 4. It was found that 114 events had some impact either in the built environment and/or in human communities and/or in the natural environment, regardless of the extent and the type(s) of impact. The many different types of impact were summarized in the next main types: human deaths (casualties) (D), damage in vessels (VD), damage in buildings and other engineered structures (SD), land impact (LI), and environmental impact (EI). Land impact refers to soil erosion and other geomorphological changes caused by the tsunamis, to destruction of cultivated land due to tsunami inundation as well as to the simple land flooding. On the other hand, we considered that the environmental impact may include a variety of effects, such as fishes and shells left behind after the tsunami, changes in the sea color, death of animals, boulder replacement from offshore to onshore, as well as tsunami sediment deposition.

One difficulty that we faced in classifying the several types of impact was that the historical descriptions often are not precise enough to decide

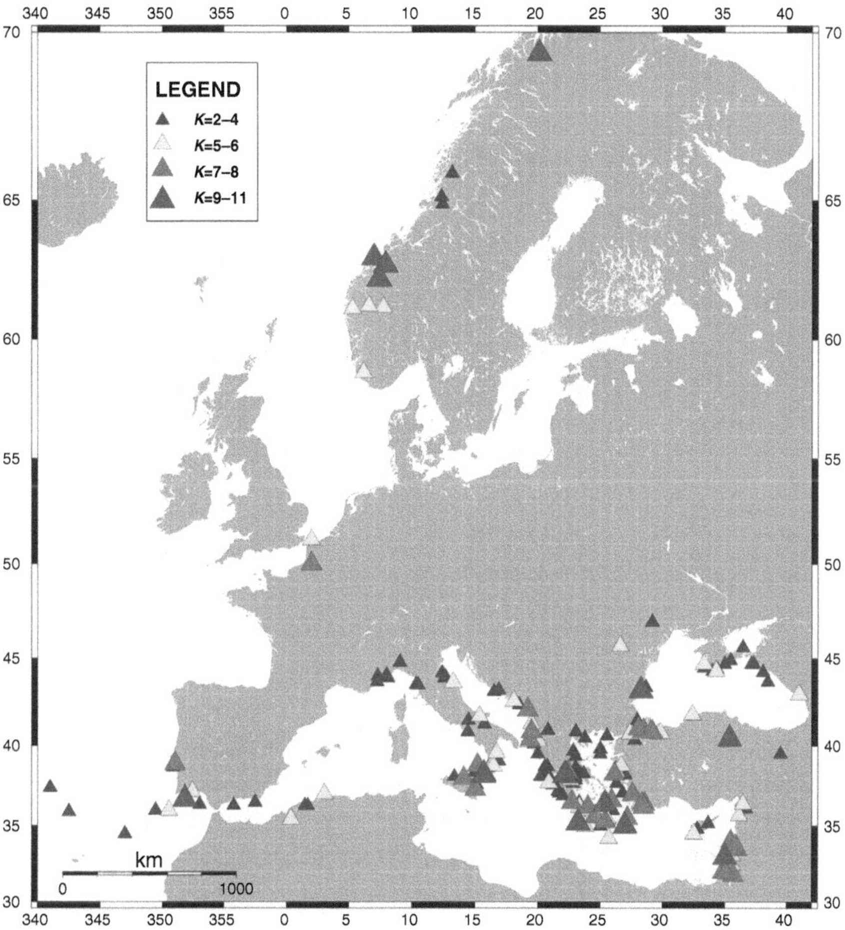

Figure 3.10 ***Geographical distribution of the tsunami sources.*** *K* is the maximum intensity in the 12-point Papadopoulos and Imamura (2001) scale.

if the impact was of this or that type. For example, for the large tsunami of AD 365 produced by a magnitude 8+ earthquake in the Hellenic Arc, among others it was reported that "humans were affected". However, such a description does not provide the information needed to conclude of what type the tsunami impact was.

The term "impact" used so far is quite general, expressing the entire spectrum of impact extent, from the very low to the highest one. Therefore, we classified the tsunami impact in four levels. Level 1 corresponds to the lowest degree of impact while level 4 corresponds to the highest degree. For example, the statement "one person was killed" means impact

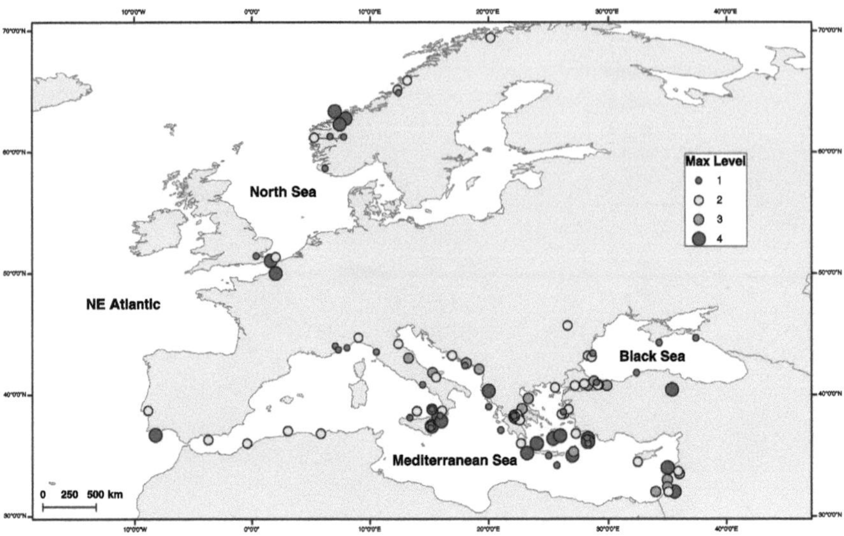

Figure 3.11 ***Distribution of maximum impact level of tsunamis.***

type D of level 1 (D1), while the statement "more than 20 persons were perished" is translated to mean D4. Figure 3.11 shows the maximum impact level per tsunami event regardless of the type of impact. Following the same geographical distribution with the distribution of tsunami intensity (Figure 3.10), the maximum impact is concentrated in the eastern Mediterranean basin, primarily along the Hellenic Arc and secondary in South Italy and in the Levantine Sea in the easternmost side of the Mediterranean Sea. However, one may not underestimate that past tsunamis had some impact also in South West Iberia, in the Strait of Dover and in western Norway.

The statistics per impact type and level are represented in Figures 3.12–3.16. One may see that, as it was expected, the majority of the events that had some impact, regardless the type(s) and the level of impact, are concentrated in the eastern Mediterranean basin and particularly in the area of the Hellenic Arc. The most common type of impact reported was on vessels (65 events) (Figure 3.12). The impact on structures was reported in 40 events (Figure 3.13). A number of 34 lethal tsunami events were reported (Figure 3.14), while the events that had land and environmental impact were 28 (Figure 3.15) and 27 (Figure 3.16), respectively.

The examination of individual tsunami events showed that most significant impact comes from extreme tsunamis caused by earthquakes of magnitude around 8 or more, such as the ones of AD 365 (Figure 3.17) and 1303

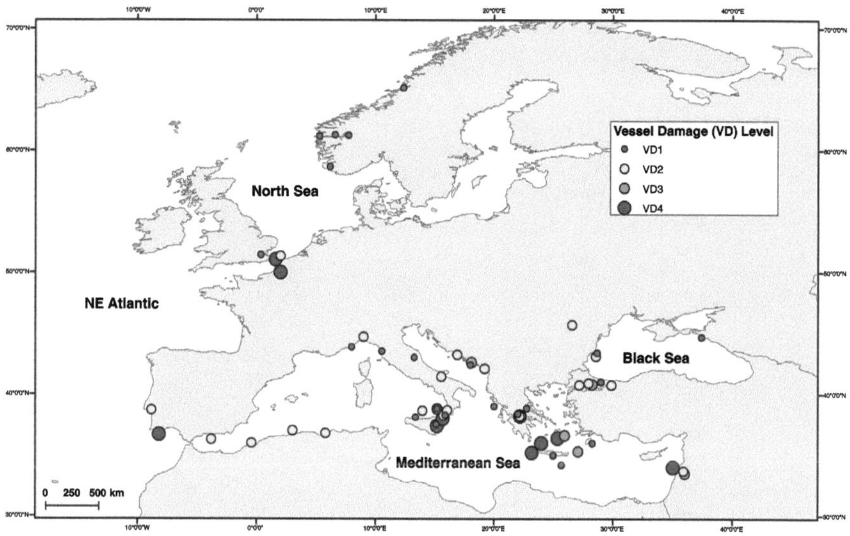

Figure 3.12 *Distribution of tsunami impact on vessels.*

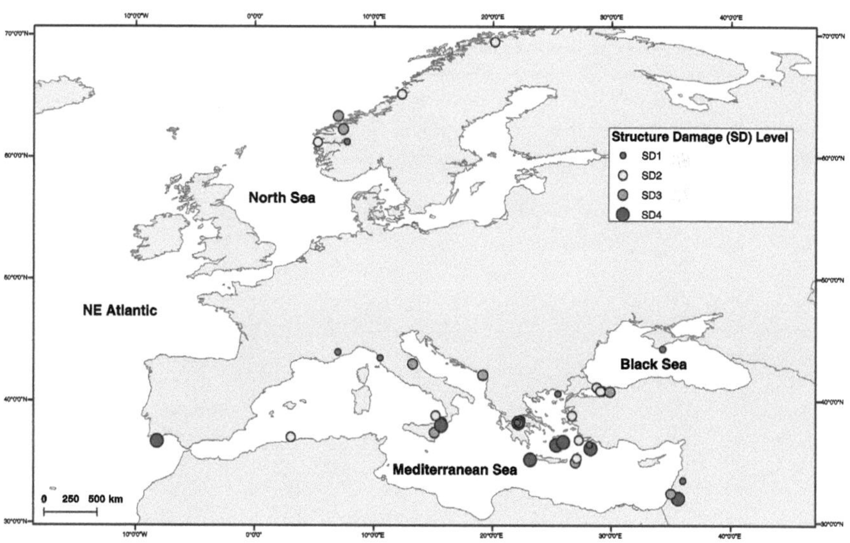

Figure 3.13 *Distribution of tsunami impact on engineered structures.*

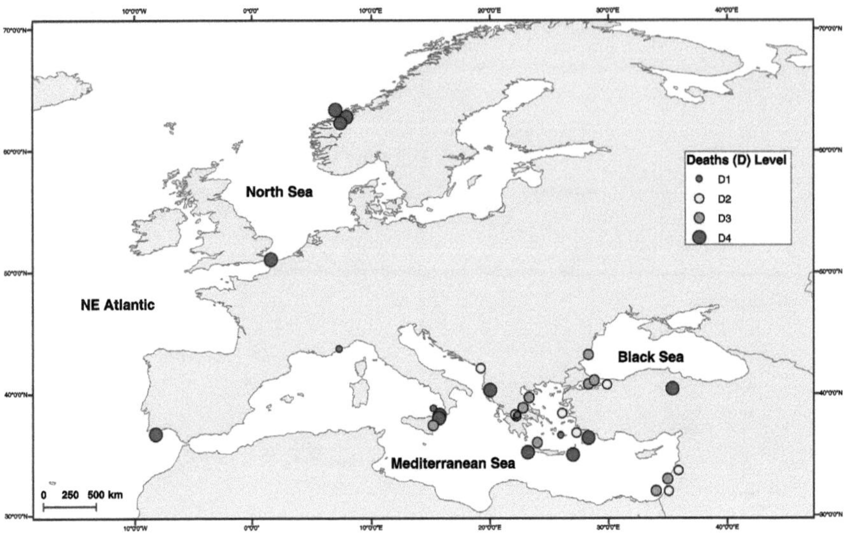

Figure 3.14 *Distribution of lethal tsunamis.*

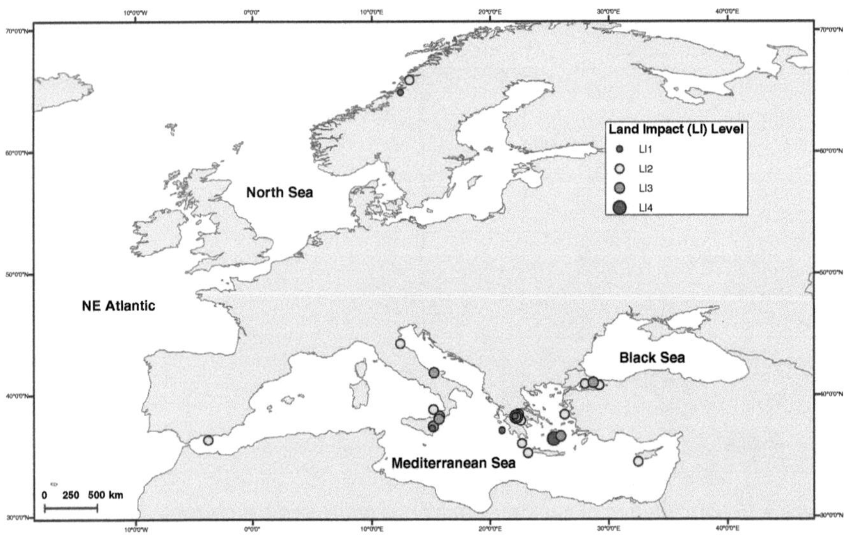

Figure 3.15 *Distribution of tsunami land impact.*

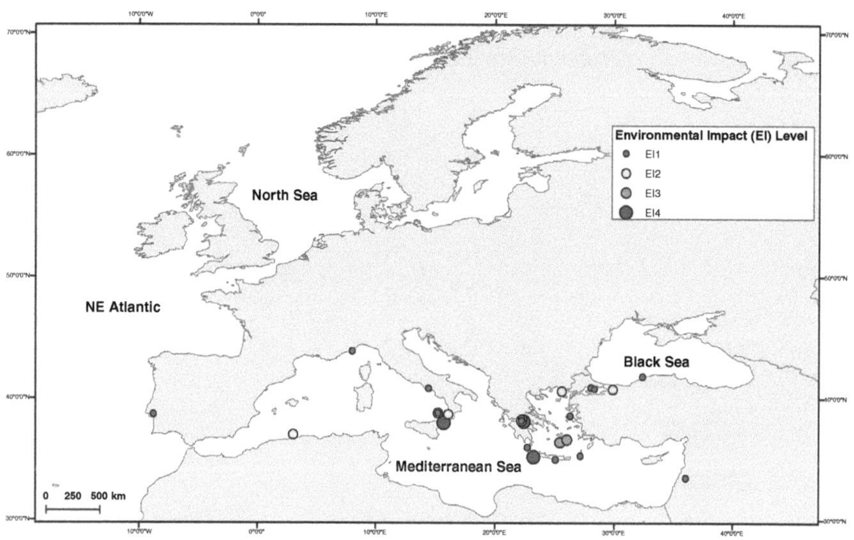

Figure 3.16 ***Distribution of tsunami environmental impact.***

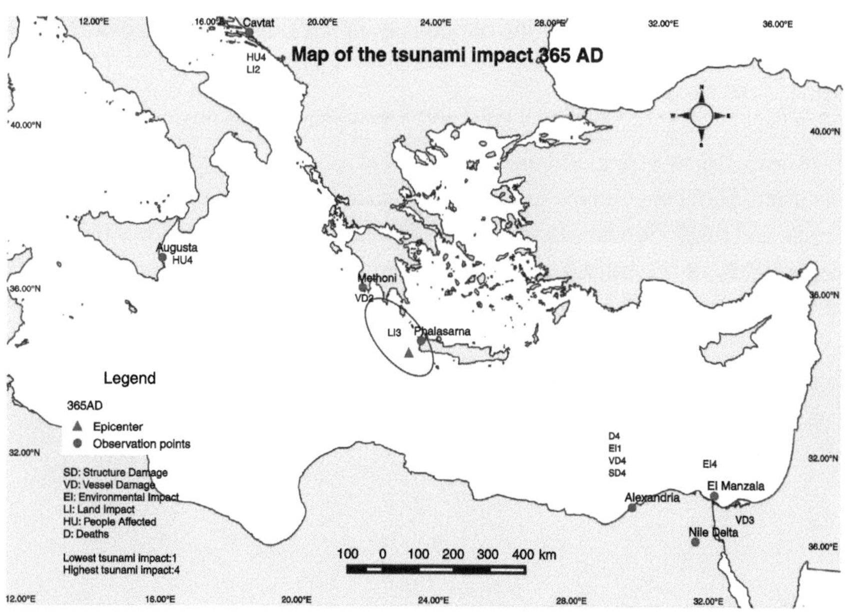

Figure 3.17 ***Geographical distribution of the coastal spots, which were reportedly affected by the AD July 21, 365 tsunami generated in the western segment of the Hellenic Arc.*** The ellipse to the west of Crete Island shows the suggested rupture zone of the causative earthquake.

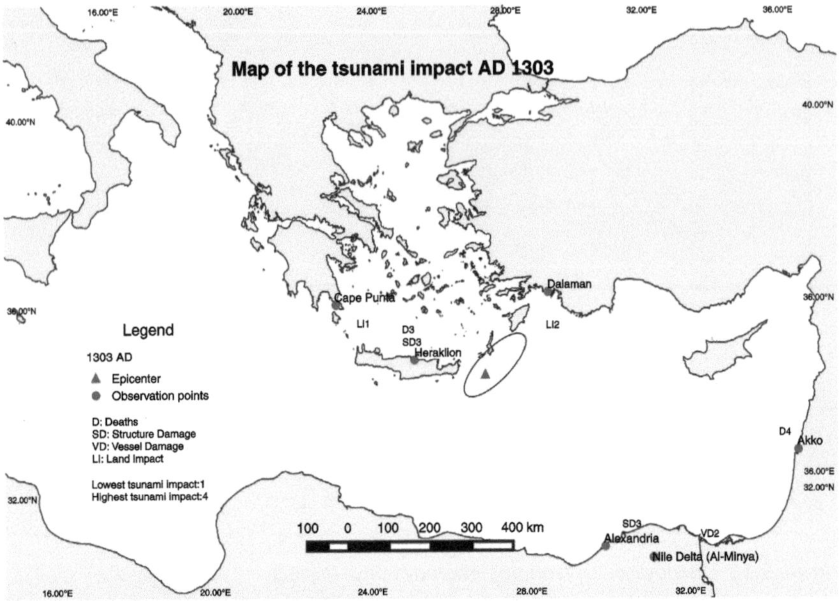

Figure 3.18 ***Geographical distribution of the coastal spots, which were reportedly affected by the AD August 8, 1303 tsunami generated in the eastern segment of the Hellenic Arc.*** The ellipse to the west of Crete Island shows the suggested rupture zone of the causative earthquake.

(Figure 3.18) in Crete and the 1755 in Lisbon. However, high impact also occurred from events of lower magnitude, such as the 1908 tsunami in Messina Straits and the 1956 tsunami in the South Aegean, which underlines the strong dependence of the impact on the community exposure (Chapter 6). Another important finding is that the cumulative impact of relatively moderate or even small, local tsunamis, produced by earthquakes, landslides, or volcanic activity, is quite important and that such distributed tsunami impact should not be neglected in actions undertaken for the tsunami-risk mitigation.

CHAPTER 4

Tsunamigenic Sources and Generation Mechanisms

4.1 INTRODUCTION

The various tsunamigenic sources existing in the European and Mediterranean region can be classified according to their geographic zonation, their potential for tsunami generation and their generation mechanisms. As regards tsunami zonation, in an earlier effort to determine the geography of historical tsunamis in Greece and the surrounding regions, Papadopoulos and Chalkis (1984) were able to determine ten coastal areas which are the most prone to experience tsunamis in the Aegean Sea, the Ionian Sea, the Marmara Sea, and offshore Albania. Later on, maps of tsunamigenic zones covering the entire Mediterranean and its connected seas were published (Soloviev et al., 2000; Papadopoulos and Fokaefs, 2005; Papadopoulos, 2009). Each one of these zones was classified in a relative scale of potential for tsunami generation based on the event size, expressed in terms of intensity, and the frequency of the historical tsunami events.

More recently, this approach was extended by including additional tsunamigenic zones in the Black Sea as well as offshore SW Iberia in the North East Atlantic Ocean (Papadopoulos et al., 2014a, see also Figure 2.1). In the tectonic rift of Corinth Gulf (zone 12 in Figure 2.1), the very high frequency of tsunami generation, regardless the tsunami size and generation mechanisms (Papadopoulos, 2003b), indicates that this rift is placed at the top of the scale with very high tsunami potential. However, tsunami waves which are produced within Corinth Gulf are only local and incapable to propagate outside the gulf, therefore, they are not the largest in the Mediterranean Sea. On the contrary, large basin-wide tsunamis were reported historically to have been generated by very large earthquakes occurring in the eastern Mediterranean Sea basin, particularly in the Hellenic Arc, such as those of AD 365 and 1303, as well as offshore SW Iberia such as the AD 1755 one.

Tsunamis in the European-Mediterranean Region. http://dx.doi.org/10.1016/B978-0-12-420224-5.00004-1

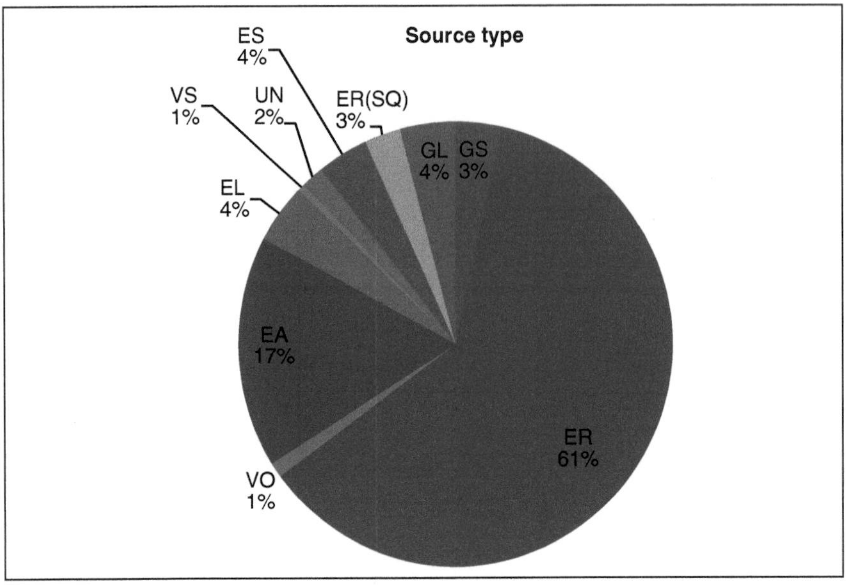

Figure 4.1 ***Percentage frequency of the several types of tsunami sources observed in the European and Mediterranean region.*** The statistics are based on the new tsunami catalog compiled by Diakogianni et al. (2015) as explained in Chapter 2. Symbol key: ER, earthquake; EA, earthquake associated; EL, earthquake landslide; ES, earthquake marine slide; VO, volcanic activity; VS, volcanic marine slide; UN, unknown; GL, gravity landslide; GS, gravity marine slide. The symbol ER(SQ) means seaquake. Seaquakes were also inserted in the catalog with the lowest tsunami reliability, that is 1, with the purpose to stand there for possible further verification. Tsunami reliability is measured in a scale from 1 to 4.

4.2 TSUNAMI SOURCES AND GENERATION MECHANISMS

In the European and Mediterranean region a variety of tsunami generation mechanisms can be recognized. Following the classification presented in Chapter 1 (Table 1.2), all types of *seismic* (or tectonic) and *nonseismic* types of tsunami sources can be found (Figure 4.1). A seismic mechanism implies that the tsunami is generated by the coseismic fault dislocation. Otherwise the mechanism is nonseismic and may include either *aseismic* tsunamis caused by landslides, volcanic activity (e.g., pyroclastic flows, volcanic landslides, cone collapse) or other causes without the involvement of earthquake activity, or *pseudoseismic* tsunamis produced by landslides, rock falls, and volcano collapses due to earthquake activity which acts only as a triggering agent. Mechanisms involving seismic and aseismic components should not be ruled out.

Figure 4.2 ***General view of the uplifted ancient harbor of Phalasarna.*** *(Photo courtesy by G. A. Papadopoulos.)*

In the next part of this chapter, examples of generation mechanisms proposed for some important tsunamis are discussed with the aim to emphasize the difficulties that the scientific community faces when characterizing the tsunami source types and generation mechanisms in the European and Mediterranean region. Also, several geophysical and empirical tools which are in use to discriminate between different source types and mechanisms are particularly discussed.

Two of the most important tsunamis in the Mediterranean regions have been the ones of December 28, 1908 in the Messina Straits, South Italy, and of July 9, 1956 in the Cyclades island complex, South Aegean Sea, Greece. Both were generated by strong tectonic earthquakes measuring magnitudes of 7.2 and 7.5, respectively (see Chapter 2 for more details). The importance of these two tsunami events relies not only on the important catastrophic results that they produced (see Chapter 3), but also that their generation mechanisms are still unclear. For this reason it was preferred to examine these two events in Section 4.3 where several criteria for the discrimination of the tsunami sources are analyzed.

4.2.1 Seismic Tsunamis

4.2.1.1 Western Hellenic Arc

Although the majority of tsunamis is generated in association with earthquake activity, only for a very limited number of events we may assume the generation mechanism. For example, in the Hellenic Arc a characteristic case of subduction-related seismic tsunami was the one of AD July 21 365 that ruptured at the western segment of the arc. Coastal tectonic elevations with amplitude decreasing gradually from SSE to NNW, measured up to 9 m in western Crete and 3 m in Antikythira Isl., were ^{14}C dated and calibrated to calendar ages of AD 341–439 and AD 265–491, respectively (Thommeret et al., 1981; Pirazzoli et al., 1992). On the basis of this tectonic evidence, the 365 event has been considered as the causative event which is responsible for the tremendous uplift which raised the Hellenistic/Roman harbor of Phalasarna in NW Crete by ~6.6 m. In the uplifted harbor of Phalasarna (Figure 4.2), Pirazzoli et al. (1992) described tsunami sediment deposits attributed to the 365 wave. However, Dominey-Howes et al. (1998) supported that the sedimentary record contradicts the tectonic interpretation. On the other hand, Shaw et al. (2008) argued that the 365 large earthquake was originated not on the subduction interface beneath Crete, but on a splay fault with a dip of c. 30° within the overriding plate.

The conclusion is that today little doubt remains that the 365 tsunami was generated by coseismic fault displacement in the western segment of the Hellenic Arc, regardless which fault was activated and if the tsunami sedimentary record is clear enough or not (see extensive review in Papadopoulos, 2011).

4.2.1.2 Eastern Hellenic Arc

In the eastern segment of the Hellenic Arc, however, a similar coseismic tectonic displacement was not documented so far in relation to the AD August 8, 1303 very large tsunamigenic earthquake. The investigation of such a tectonic signature between Crete and Rhodes islands, either onshore or offshore, constitutes an important challenge for better understanding the tsunami generation mechanism in that tectonic segment. Further to the easternmost segment of the Hellenic Arc, that is, in the Dodecanese islands and in SW Turkey, where the seismotectonic setting is complicated, historical seismic tsunamis (e.g., AD 1481, 1609, 1741) were reported to have been generated possibly in the Rhodes Abyssal Plain where dip-slip tectonics prevails and the sea depth is around 3 km (see Papadopoulos et al., 2007a, 2012a and references therein). However, the generation mechanisms of those tsunamis still remain unidentified.

4.2.1.3 Ionian Sea and North Aegean Sea

In areas of high seismicity which are dominated by active strike-slip faulting, such as in the central Ionian Sea and the North Aegean Sea trough, the tsunami activity is significantly low. In fact, in the central Ionian Sea area between tsunamigenic zones 18 and 19 (Figure 2.1), which is dominated by the Cephalonia-Lefkada transform fault, no tsunami generation has been historically reported although the seismicity is very high there (e.g., the large *M*7.2 earthquake of August 12, 1953). In the North Aegean Sea (zone 15 in Figure 2.1), very low tsunami activity has been historically reported (Papadopoulos, 2009). On the contrary, large magnitude earthquakes occur frequently there, for example, the earthquakes of February 19, 1968 (*M*7.1), December 19, 1981 (*M*7.2) and January 18, 1982 (*M*7.0) (Papadopoulos et al., 2002).

4.2.1.4 Marmara Sea

In the Marmara Sea it is likely that the predominant tsunami generation mechanism is the earthquake activity and associated coastal and submarine landslides (Yalciner et al., 2002; Minoura et al., 2005). A characteristic example is the *M*7.9 earthquake of AD December 26, 1939, which ruptured a long segment of the North Anatolian Fault at a distance of ~100 km inland from the Turkish Black Sea coast. After this very large earthquake a tsunami was observed in Black Sea coastal localities offshore north Turkey. The wave was recorded by Soviet tide gauges in the eastern Black Sea. A solution regarding the generation mechanism is the one which involves coseismic landsliding at the continental slope of the Black Sea between Sinop and Batumi (Papadopoulos et al., 2011). One may suggest that a secondary fault was activated with the main earthquake. However, seismic records do not support the case, although we should take into account that the seismograph coverage of the area was very poor.

4.2.1.5 SW Iberia

The source type and the generation mechanism of the AD November 1, 1755 Lisbon large earthquake and tsunami has been studied by several authors on the basis of a variety of geological evidence, geophysical data, and tsunami modeling. However, many different geodynamic models and mechanisms have been proposed (Gutscher et al., 2002; Baptista et al., 2003; Gràcia et al., 2003a,b; Terrinha et al., 2003, 2009; Thiebot and Gutscher, 2006; Zitellini et al., 2004, 2009; Stich et al., 2007). None of these models, however, satisfactorily accounts for the estimated earthquake magnitude and the tsunami arrival times at the different localities onshore (Gràcia et al., 2010). Other relevant structures are the long WNW–ESE strike–slip faults known as the SWIM lineaments that extend hundreds of km from the Horseshoe Abyssal Plain to the inner part of the Gulf of Cádiz (Zitellini et al., 2009; Terrinha et al., 2009; Bartolome et al., 2012; Martínez-Loriente et al., 2013) (Figure 2.25). These structures correspond to large strike-slip faults, involving only small vertical component. Consequently, they are likely not the best candidate tsunamigenic structures to interpret the 1755 tsunami generation.

The examples analyzed above clearly indicate that even for large magnitude earthquakes the characterization of the tsunamigenic seismic source is not an easy task due to the lack of adequate data. In view of this, a common practice followed by authors who try to numerically simulate such tsunami events is to insert many assumptions regarding the type of source and the magnitude of the causative earthquake. For example, for the simulation of the May 11, 1222 tsunami in SW Cyprus a very likely exaggerated earthquake magnitude of ~7.5 was adopted (Yolsal et al., 2007).

4.2.2 Landslide Tsunamis

Landslide tsunamis are produced either by the action of gravity in sediments of reduced shear strength (e.g., due to high pore pressure) without any external triggering force or by geodynamic processes such as strong earthquakes and volcanic eruptions that may set in motion unstable rock and/or soil masses. The geographic distribution of landslide tsunami sources historically documented in the Mediterranean region is illustrated in Figure 4.3.

4.2.2.1 Aseismic Landslide Tsunamis

Typical cases of tsunamis produced by aseismic landslides were the ones of AD February 7, 1963 in the western side of Corinth Gulf, Central Greece, and of AD October 16, 1979 in Nice, Côte d'Azur, France. Since tsunamis of this type are local but take large amplitude in the near-field domain, both waves caused victims and significant destruction in near-source coastal segments.

A variety of analytical models are available for the generation of tsunamis by submarine landslides. The 1963 Corinth Gulf tsunami is a key event and, therefore, relevant numerical modeling results are examined later in the next chapter. It is of interest to note that in the western Corinth Gulf, extensive submergence of a coastal strip, which was caused by the M~6.6 earthquake of AD December 26, 1861, very likely generated the tsunami

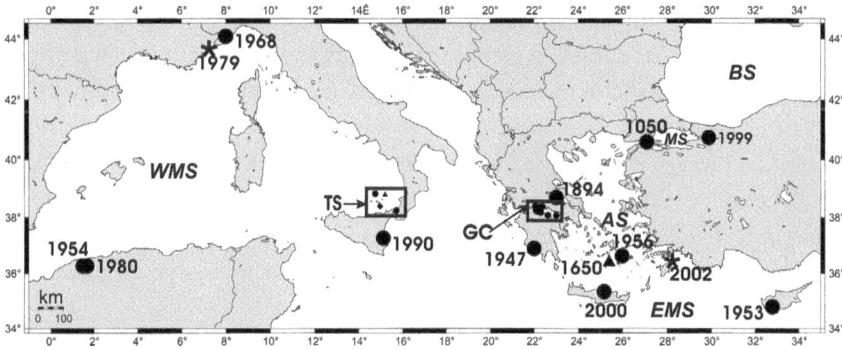

Figure 4.3 ***The geographic distribution of landslide tsunami sources historically documented in the Mediterranean region (After Papadopoulos et al., 2007b, 2014a.)*** Figures indicate years of occurrence. Key for geography: EMS, East Mediterranean; WMS, West Mediterranean, MS, Marmara Sea; BS, Black Sea; AS, Aegean Sea; GC, Gulf of Corinth; TS, Tyrrhenian Sea. Symbol key: solid circle, earthquake landslide; solid triangle, volcanic landslide; asterisk, gravity landslide.

that was reportedly observed after the earthquake (Papadopoulos, 2003b, and references therein).

In the Ligurian Sea, on October 16, 1979 an aseismic slope failure event occurred at the very narrow and steep continental slope offshore Nice, involving a portion of the Var River delta and runaway fill of the Nice airport extension which was under construction at the time (Gennesseaux et al., 1980; Dan et al., 2007; Sahal and Lemahieu, 2011). The near-field wave heights were successfully simulated (Assier-Rzadkiewicz et al., 2000). However, the theoretical results were not in complete agreement with the far-field observations. This is explainable by the rapid amplitude attenuation due to strong wave dispersion, a common feature of landslide-generated tsunamis (Papadopoulos and Kortekaas, 2003).

In western Norway several tsunami events have been historically documented (see Chapter 2). All of them were caused by landslides and rockfalls.

4.2.2.2 Landslide Tsunamis Caused by Earthquakes

In the Levantine Sea, including the area of Cyprus island, that is from Egypt in the south to SE Turkey in the north, about 24 tsunamis have been reported during the last 3.5 millennia. The majority of these tsunamis were reported to occur in the last two millennia (Fokaefs and Papadopoulos, 2007; Salamon et al., 2007, 2009), which is evidence of the vast incompleteness of tsunami data in the early historical period. About two thirds of these tsunamis were local in stretch, but a third were basin-wide. However, the generation of locally strong tsunamis by earthquakes occurring inland along the strike-slip Levantine rift lacks a reasonable interpretation. Interestingly, most of the abnormal tsunamis and sea-storms reported in history to affect the Levantine coast were said to be associated with earthquakes – now recognized to have ruptured inland. This unexpected association attracted much interest for how inland earthquakes are capable of triggering a tsunami (see Salamon et al., 2007 and references therein). The missing link was suggested to be submarine landslides, for numerous typical landslide scars exist along continental slopes of the Levantine Sea.

The case of the tsunamigenic earthquake of AD May 1, 1202 (Figure 2.9) is an excellent example: There ... *Gigantic waves rose up in the sea between Cyprus and the coast of Syria...* (summarized by Guidoboni and Comastri, 2005). Damage reports (Ambraseys and Melville, 1988) and palaeoseismic studies have shown important seismic ruptures along the faults of the Yammouneh (Daëron et al., 2005; Nemer et al., 2008) and of the Jordan Gorge (Marco et al., 1997, 2005; Ellenblum et al., 1998) belonging to the system of the Dead Sea Transform Fault (DSTF). The hypothesis of submarine landslides is supported by more recent observations in similar seismotectonic settings, such as the large, continental *M*7.9 earthquake of AD December 26, 1939 that caused a tsunami in the Black Sea although it is documented that the earthquake ruptured along the North Anatolian Fault about 100 km inland.

A good historical example of a tsunami triggered by coseismic landslide was the one of AD February 6, 1783 in Calabria, south Italy (Figure 2.15). Such a tsunami generation mechanism was initially suggested on the basis of the information contained in documentary sources (Graziani et al., 2006). Later, this hypothesis was verified by subaerial and submarine geological surveys performed in the area (Bozzano et al., 2011). In the western Mediterranean and particularly in the North Algerian margin, well-known historical examples of tsunamis produced by submarine landslides were the ones triggered by the large earthquakes of AD September 9, 1954 and October 10, 1980 (El Robrini et al., 1985; Soloviev et al., 2000) (see zone 1 in Figure 2.1).

A multidisciplinary marine geological and geophysical dataset acquired during the last years offshore the Iberian Peninsula in the frame of several research projects, revealed a number of slope failures in the South and East Iberian margins which may represent a tsunami hazard for the coasts of Portugal, Spain and North Africa (Baraza et al., 1990; Rothwell et al., 1998; Lastras et al., 2002, 2004, 2007; Gràcia et al., 2003a,b, 2006, 2010; Terrinha et al., 2003, 2009; Droz et al., 2006; Urgeles et al., 2007; Vizcaino et al., 2006;

Camerlenghi et al., 2009; Zitellini et al., 2009; Cattaneo et al., 2010). A good example is the newly discovered North Gorringe Avalanche (NGA), which is a large (~80 km^3 and 35 km runout), deep water (2900 m to 5100 m depth) mass failure situated at the northern flank of Gorringe Bank in the southwest Iberian margin. Near-field tsunami simulations showed that a mass failure similar to the NGA could generate a tsunami wave of more than 15 m high that would hit the south Portuguese coasts in ~30 min after its generation. These results imply that deep-water landslides require more attention as potential tsunami sources in SW Europe (Lo Iacono et al., 2012).

4.2.3 Volcanic Tsunamis

Volcanic eruptions are much less frequent than the frequency of earthquakes in the Mediterranean Sea and in the North East Atlantic. Although only a few tsunamis were attributed to volcanic activity, at least three cases are of particular value, all of them occurring in the Mediterranean Sea. Two of the largest tsunamis known in the Mediterranean Sea were generated by strong eruptions taking place in the Thera volcanic island complex: the LBA (seventeenth century BC) Thera tsunami and the AD September 29, 1650 Columbo tsunami. Local but strong tsunamis were repeatedly reported to have been caused by landslides triggered by the eruptive activity of Stromboli, the last one occurring on AD December 30, 2002. However, with the exception of the last event, which is discussed later as a good example of volcanic landslide tsunami, the generation mechanisms of the two tsunamis in Thera and Columbo are not precisely known. Numerical modeling results are helpful in this direction and, therefore, they are examined in Chapter 5.

4.2.3.1 Volcanic Landslide Tsunamis

Evidence for massive debris avalanches from Mt. Etna which entered the Ionian Sea in the early Holocene was interpreted by Pareschi et al. (2006a) to have been responsible for tsunami generation and subsequent destructions in coastal communities. Namely, the above authors suggested that Mt. Etna's massive debris avalanches generated a large tsunami impacting all of the eastern Mediterranean Sea. As a result of the tsunami action deposits of homogenites were formatted in the Ionian Sea (Pareschi et al., 2006b) while the Neolithic village Atlit-Yam in the Israeli coast was destroyed (Pareschi et al., 2007, 2008). Although no criticism appeared as regards the tsunami generation *per se*, Galili et al. (2008) showed that the tsunami destruction in Atlit-Yam finds no support in the archeological, anthropological, faunal, botanical, or sedimentary record in the site, which instead was abandoned due to a postglacial rise of the sea level.

One of the best cases of landslide-induced tsunamis during volcanic activity is the one of the Stromboli volcano which occurred during the eruptive episode of AD December 30, 2002. Eyewitness accounts (Tinti et al., 2005c,d) and impact observations performed during post-event field surveys (Tinti et al., 2006a) agree with the results of numerical simulations (Tinti et al., 2006b; Maramai et al., 2005b) that two landslides occurring with a time difference of about 7 min caused two corresponding damaging tsunamis which took maximum wave height of about 9–10 m and caused extensive indoor and outdoor destruction in the coastal villages of Stromboli (Figures 2.16 and 2.17). A multibeam bathymetry down to 1 km of depth collected just 10 months before the event and repeated after it showed how preexisting features interacted with the slide event in controlling the instability (Chiocci et al., 2008a,b). Seismic signals from broadband and short-period seismic stations showed that the landslides involved both the submarine and the subaerial northwest flank of the volcano with the process starting from the submarine section (La Rocca et al., 2004).

4.3 SOURCE DISCRIMINATION

From the examples of tsunamis examined earlier it appears that the discrimination between the different types of sources and generation mechanisms is a quite puzzling issue in the European and Mediterranean region. Characterization of a tsunami source includes the type of the source, the source mechanism responsible for tsunami generation, and the source dimensions, geometry, and kinematics. The first issue which is of relevance here is the discrimination of the tsunami sources, which as discussed earlier, is not an easy task. Whether the tsunami is of tectonic, landslide, or volcanic origin, there is an enormous difference especially in the expected effects of the event but also on its repeatability (Papadopoulos et al., 2014a). In addition, for most of the historical tsunamis which are known in the Mediterranean and in the North East Atlantic, the definition of the actual source is still a matter of debate (e.g., LBA Thera tsunami, AD 1693 and 1908 tsunamis produced by earthquakes in Sicily and South Italy, AD 1650 Kolumbo volcanic tsunami, also the earthquake-generated tsunamis of AD 1755 in SW Iberia and the AD 1956 in the South Aegean). In the forthcoming sections of this chapter, several criteria are examined as for their efficiency to discriminate tsunami sources. In particular, as regards the treatment of historical events which are not documented by instrumental records a pilot analysis is proposed taking the strong earthquake and tsunami of May 11, 1222 in SW Cyprus as a case study. However, the contribution of numerical modeling inputs for the characterization of tsunami sources is examined analytically in Chapter 5.

4.3.1 Field and Instrumental Observations

The analysis of historical or geological data describing the size, extent and location of inundations produced by tsunamis of the past can be effective to help solving this problem. For example, from the study of a set of well-documented landslide tsunamis observed around the globe, including the 1963 Corinth Gulf, 1979 Nice and 1999 Izmit tsunamis (Figures 4.4 and 4.5), it was found that slump volume seems to control both the maximum wave height and maximum length of affected coastline (Papadopoulos and Kortekaas, 2003). A rapid, quasi-exponential attenuation of wave heights with distance from the source was observed due to strong wave dispersion. Therefore, such empirical relationships are useful to discriminate landslides from other types of tsunami sources.

An approach introduced by Okal and Synolakis (2004) is based on the observation that the distribution of high runups and the extent of the inundated coast are very different in the case of the occurrence of a localized landslide, a gigantic collapse, or a displacement of the sea/ocean bottom. Such criteria were tested, for example, in Sicily tsunamis of AD 1693, 1783, and 1908 but some contradictory results were obtained. In fact, Gerardi et al. (2008) concluded that the 1693 and 1908 tsunamis were caused by seismic dislocation sources while the 1783 one was due to earthquake-triggered submarine landslide. By contrast, Billi et al. (2010) found that the aforementioned three tsunami events were produced by submarine landslides. The scenario can be even more complex considering that seafloor displacement and gravitational failure can be concurrent.

A typical historical case of a strong earthquake-generated tsunami whose causative fault remains unidentified is the one of AD July 30, 1627 in the Gargano promontory in Apulia, Italian coast of Adriatic Sea. The Gargano 1627 tsunami was quite large according to reliable though scarce historical sources (Tinti et al., 2004), but the causative fault is still far from being identified, due to the absence of significant dip-slip faults in the area (Tinti et al., 1995; Tinti and Piatanesi, 1996). The local tectonics is rather complex and thought unfavorable to tsunami generation since strike-slip faults are the

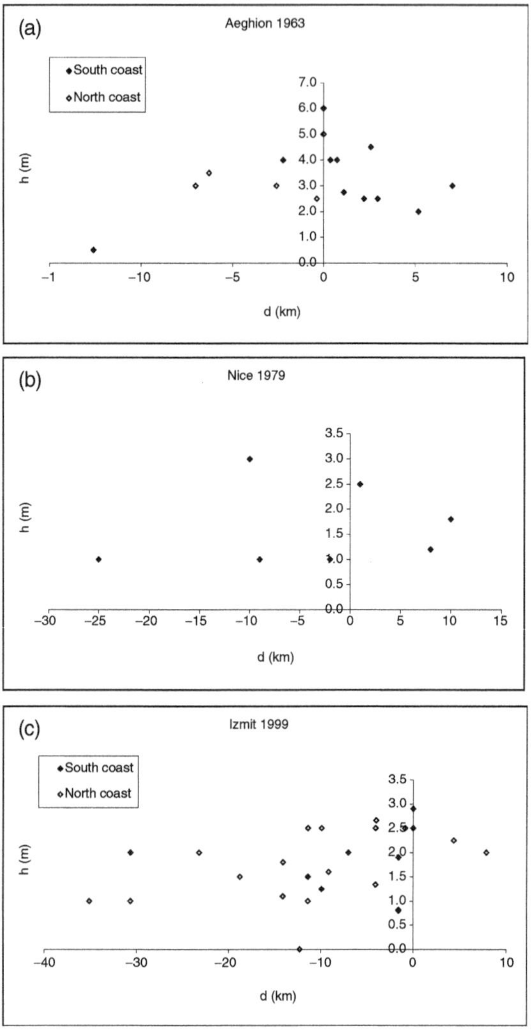

Figure 4.4 ***Tsunami wave heights,* h *(m), observed at distance* d *(km) from the tsunami source for three landslide-triggered tsunami events reported (a) on February 7, 1963 in Aeghion, west Corinth Gulf, Central Greece, (b) on October 16, 1979 in Nice, Côte d'Azur, France, and (c) on August 17, 1999 in Izmit Bay, Marmara Sea, Turkey (After Papadopoulos and Kortekaas, 2003.)*** The two first events were aseismic while the third landslide was triggered by a large *M*7.5 earthquake.

predominant feature north and south of the promontory. The most relevant one is the Mattinata fault, striking east–west south of Gargano. This fault is considered as being still active as shown by a number of sound evidence, such as field morphology, InSAR, and GPS data, pointed out also by recent studies

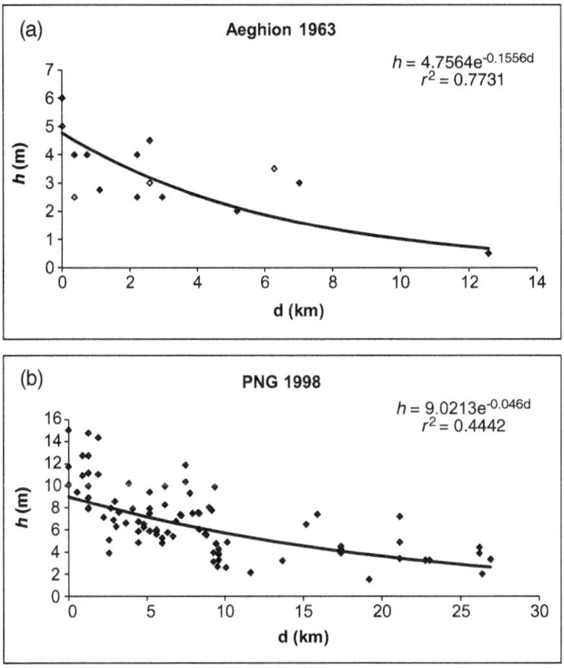

Figure 4.5 ***Attenuation of tsunami wave heights,*** **h** ***(m), as a function of distance*** **d** ***(km) from the tsunami source for the aseismic landslide tsunami of (a) February 7, 1963 in Aeghion, west Corinth Gulf, Central Greece and of (b) Papua New Guinea (PNG) tsunami of July 17, 1998 triggered by a submarine landslide caused by a strong*** **M7.1** ***earthquake on October 16, 1998 (After Papadopoulos and Kortekaas, 2003.)***

(see Fracassi et al., 2012, and references therein). It has also been suggested that this fault may have an offshore active continuation in the Gondola fault zone.

Of particular importance is to investigate further the generation mechanisms of more recent tsunamis which still remains questionable. Characteristic examples are the tsunamis of 1908 in Messina Straits and of 1956 in Amorgos, South Aegean. For both events the collected seismological, tectonic, geodetic, and tide gauge data, as well as results from numerical tsunami simulations, proved insufficient to distinguish between seismic, landslide, or combined source mechanisms. Reviewed seismological, geological, geodetic data (Valensise and Pantosti, 1992) and tectonic stress inversion (Neri et al., 2004) indicated that the Messina 1908 M_w7.1 earthquake was associated with normal faulting. However, the seismogenic fault has not yet been identified conclusively and this has hindered attempts to develop simulation models of this major tsunami. An analogous case was the large tsunami triggered in the South Aegean after the Amorgos M_w7.5 earthquake of July 9, 1956 that very likely was associated with normal faulting in the submarine trough shaped by the islands of Amorgos, Santorini (Thera), Anafi and Astypalaea (Galanopoulos, 1957a,b; Ambraseys, 1960; Papadopoulos and Pavlides, 1992; Perissoratis and Papadopoulos, 1999; Okal et al., 2009). Both the tectonic and landslide origin of the tsunami have been proposed on the basis of seismological and marine geophysics of the area and tested through numerical simulations, as presented in Chapter 5. In Section 4.3.2, however, the slowness factor to characterize the slip in the seismic fault is examined for these two key tsunami events.

4.3.2 A Pilot Historical Study: The Case of AD 1222 Earthquake and Tsunami in SW Cyprus

In this example original historical information has been used to reconstruct the macroseismic field of the earthquake and then to infer the most likely seismotectonic scenario. On the basis of the macroseismic field features the earthquake magnitude is calculated. Then, having a seismotectonic scenario and a reliable estimation of the magnitude one may go ahead with tsunami simulation.

The very strong and lethal earthquake that shook Cyprus on May 11, 1222 is known from a variety of Greek, Latin, and Vulgar French documentary sources as well as from late thirteenth century Armenian chronicles (see reviews in Ambraseys, 1965, 2009; Christofidis, 1969; Antonopoulos, 1973; Evagelatou-Notara, 1993; Guidoboni and Comastri, 2005, and references therein). Detailed descriptions can be found in some Latin sources (translations from Latin are after Guidoboni and Comastri, 2005):

1. The *Historia Damiatina* of Oliverus Scholasticus, a chronicler of the 5th Crusade (1213-1221):
   In the year of grace 1222, in the month of May, a great earthquake occurred in Egypt, at Limassol and Nicosia and in other places in the same island [Cyprus], and especially at Paphos, to the extent that the town and the castle were completely destroyed, and men and women who were taken by surprise there at the moment of the shocks, died together, ***and the harbor was left dry,*** *and shortly afterwards some springs spouted forth there.*
2. The *Dialogus miraculorum*, Chapter 48, by the Cistercian monk Caesarius of Heisterbach, Cologne, Germany, who was contemporary and obtained information from eyewitnesses:
   In this present year 1222 since the Incarnation of Our Lord, as we are told by some of those who are from the region concerned, two towns in Cyprus collapsed

in the earthquake. Indeed, in one of those [towns], a church was shaken and collapsed as the bishop was celebrating mass at the altar, killing him and some of the congregation.

3. In Chapter 49, monk Caesarius continues with this earthquake:
...at that same period in Cyprus, at Paphos, and in many other provinces and towns, many were struck down by earthquakes... This was unheard of for the times, and when at the beginning of Lent master Oliverus [Scholasticus] was preaching at Cologne, two mountains at Paphos with a lake between them moved together, thus becoming a single mountain.

Archaeological excavations have verified that people were killed in the Frankish castle of Paphos, the so-called castle of Σαράντα Κολώνες (Saranda Kolones = Forty Pillars), built only 30 years earlier overlooking the harbor. The castle was uncovered by A.H.S. Megaw revealing piles of stones that very likely had fallen in the 1222 earthquake.

In one of the documentary sources there is evidence that the strong earthquake was likely associated by a tsunami which inundated the coastal zones of both Paphos and Limassol. In fact, Marchisio Scriba, a Genoese notary, official annalist to the Republic of Genoa from 1220 onwards, says (Guidoboni and Comastri, 2005):

...But, by a divine miracle, the earth, saddened as one can imagine by the loss (of Damietta), shook on the island of Cyprus, ***and the sea withdrew from the shore, leaving a large part of the sea bed quite dry; and Paphos and Limassol were completely submerged by this sea-wave.***

This is consistent with the narration of Oliverus Scholasticus who says that *the harbor was left dry.* The castle of Paphos was never rebuilt, as it was no longer needed to protect the port, which had dried up (Rosser, 1985; Pryor, 2006). Pryor (2006) went a step further and supposed that the earthquake permanently changed Paphos, rendering the harbor unusable, and moving the shoreline seawards; it no longer had a protected anchorage. That *the sea withdrew from the shore* before Paphos and Limassol were temporarily submerged due to tsunami flooding, perhaps is due to that dip–slip coseismic fault movement occurred offshore Paphos. If the port dried up permanently, as Rosser (1985) and Pryor (2006) suggested, then ground uplift along the sea-shore may have accompanied the earthquake. The imminent result was the triggering of a local tsunami which shortly afterwards inundated both Paphos and Limassol.

The extensive destruction with many human victims caused in Paphos implies that seismic intensity of IX or X should be assigned there. Although no details can be found in the documentary sources, the earthquake very likely was damaging at Limassol as well. In fact, one source says that the earthquake occurred at Limassol, while another points that two towns in Cyprus collapsed in the earthquake. I suppose that one town was Paphos and the other was Limassol. Intensity is hardly estimated for Limassol due to the lack of sufficient information. This is also the case for Nicosia where, according to one source, the earthquake occurred also there. Very likely it was only felt, the damage produced, if any, being only slight.

That the main zone of destruction, the coseismic ground failures (landslide, new water springs), the main tsunami manifestation and the possible ground uplift, all were observed in the area of Paphos, makes it reasonable to assume that the earthquake source was very close to that city. I believe that the seismic fault, possibly a reverse one, was of about E–W in strike and dipping to north, that is its hanging wall extended from the sea-shore inland. Such a seismotectonic scenario fits well not only the concentration of the main macroseismic effects in the supposed hanging-wall domain of the fault which occupies the area of Paphos, but also the suggested permanent dryness of the port due to ground uplift and the subsequent tsunami generation.

Using the empirical relationships between earthquake magnitude and seismic intensity proposed by Papadopoulos (2011), for seismic intensity IX or X in Paphos we get earthquake magnitude $M = 6.37$ or $M = 6.62$, respectively, that is $M = 6.5$ on average. Limassol is situated at epicentral distance of about 60 km. If one adopts intensity as high as VII we get $M = 6.7$. For Nicosia, intensity V or VI yields magnitude $M = 6.28$ or $M = 6.64$, respectively, that is $M = 6.46$ on average. As a consequence, an earthquake magnitude estimation of $M = 6.5$ (± 0.2) is a realistic approach.

The seismotectonic scenario I reached at is consistent with the one adopted by other authors (Yolsal et al., 2007 and references therein). However, the magnitude estimated is nearly one order of magnitude less than the one ($M \sim 7.5$) introduced by Yolsal et al. (2007) with the aim to simulate the tsunami.

4.3.3 Slowness Factor of the Seimic Slip

A seismological criterion for source discrimination is the *slowness factor*, θ, which characterizes the seismic slip in the earthquake causative fault (Newman and Okal, 1998; Ebeling and Okal, 2012):

$$\theta = \log_{10}\left[\frac{E}{M_0}\right] \qquad [4.1]$$

Table 4.1 Slowness factor calculated for the Messina 1908 and Amorgos 1956 tsunamigenic earthquakes by following the method of Newman and Okal (1998)

Earthquake	M_w	Seismic moment ($M_0 \times 10^{27}$ *dyn–cm*)	Slowness factor
Nicaragua 1992	7.66	3.40	−6.51
Java 1994	7.87	5.30	−5.76
Peru 1996	7.18	2.20	−6.22
Flores Sea 1992	8.07	5.10	−4.58
PNG 1998	6.80	0.37	−5.50
Messina 1908	7.10	0.56	−5.39
Amorgos 1956	7.50	3.90	−4.68

For reasons of comparison, slowness factor was also inserted as it was calculated by those authors for some selected tsunamigenic earthquakes, including the classic tsunami earthquake of Nicaragua. Key: M_w, moment magnitude; M_0, seismic moment. One may observe that only the Nicaragua 1992 and Peru 1996 events satisfy the criterion for tsunami earthquakes. The next events, including the 1908 Messina and 1956 Amorgos, do not fall into this category of earthquakes.

This factor is determined by the energy, E, radiated by an earthquake over the low-frequency seismic moment, M_0. This approach was tested by Papadopoulos et al. (2014a) with fault parameters taken from Capuano et al. (1988); Pino et al. (2000) and Amoruso et al. (2002) for the Messina Strait case: length = 43 km, width = 20 km, slip = 1.5 m; from Papadopoulos and Pavlides (1992) and Okal et al. (2009) for the case of Amorgos: length = 75 km, width = 30 km, slip = 2 m. Seismic magnitudes and moments are listed in Table 4.1. The slowness factor obtained is $\theta = -5.39$ and $\theta = -4.68$ for the 1908 and 1956 earthquakes, respectively. This result implies that both earthquakes do not exhibit the exceptionally slow source behavior ($\theta \leq -6.0$) expected for *tsunami earthquakes* in the terminology introduced by Kanamori (1972) to characterize the 1992 Nicaragua earthquake source and other earthquakes whose tsunamis were disproportionately large with respect to their seismic moment (Table 4.1).

4.4 OTHER TSUNAMI SOURCES

4.4.1 Meteotsunamis

Meteorological tsunamis or meteotsunamis are barotropic ocean waves generated through resonant coupling with traveling atmospheric disturbances (e.g., Vilibić and Paklar, 2006, and references therein). Meteotsunamis are not as frequent as the tectonic tsunamis but they may cause significant flooding and damage particularly in harbors and bays. The phenomenon of meteotsunamis is known in several places of the global ocean, for example, in Japan, New Zealand, Pacific US coast, and elsewhere. In the Mediterranean Sea, it is known under the name *Rissaga* in the Balearic islands where meteotsunamis take up to 1.5 m amplitude and about a 10 min period (Monserrat et al., 1991). Such meteotsunami events occurred in Ciutadella, Balearic islands, several times in the past, for example, in 1981, 1984, 1989 (Gomis et al., 1993). Similar events have been observed in the Adriatic Sea (Vilibić and Paklar, 2006; Raicich et al., 1999) as well as in the eastern Aegean Sea (Papadopoulos, 1992) and in the Black Sea (Dotsenko and Miklashevskaya, 2008; Vilibić et al., 2010).

The coastal behavior of meteotsunamis is similar to that of tsunamis. Marcos et al. (2003) noted that in some occasions, for example, in the Balearic islands, these oscillations are stronger than expected, even taking into account the large amplification factor by resonance of the inlet. Those authors showed numerically how the phase speed of the atmospheric pressure disturbance generating the surface waves is a critical factor in the energy transfer between the atmosphere and the ocean.

4.4.2 Tsunamis From Asteroid Impacts

Asteroid impacts, including tsunami generation, have been geologically documented in several geological periods and places on the Earth (see review in Kharif and Pelinovsky, 2005). Calculations have shown that the magnitude of the earthquake induced by a 0.5 km size asteroid is larger than 8, and the tsunami wave may be 50 m or higher at epicentral distance of 1000 km from the impact point (Koshiishi, 1998).

There is no stable geological evidence for tsunami generation from asteroid impact in the European-Mediterranean region. However, some modeling results regarding tsunami forecasting from asteroid impact have been obtained. For example, Ward and Asphaug (2003) modeled the possible collision of a 1.1 km diameter asteroid (1950 DA) with the Earth in the year AD 2880. Traveling with a speed of 17.8 km/s the asteroid may strike the ocean 600 km east of the United States coast and would blow a cavity 19 km in diameter. The use of the linear theory for a fluid of constant depth together with shoaling and runup corrections allows the prediction of tsunami heights along the coast. Waves would strike the coast of Europe and Africa within 12 h. The wave height would reach 23 m in Ireland, 16 m in the United Kingdom, 17–21 m in France, and 15 m in Portugal. However, the detailed relief of the seafloor was not considered and, therefore, effects of refraction and diffraction were not taken into account in these calculations (Kharif and Pelinovsky, 2005). As a consequence, such calculations may underestimate or overestimate the values of tsunami heights obtained.

In the Black Sea, significant beach runup of impact-generated waves were calculated but it was concluded that the potential emissions of toxic and flammable gases could be more disastrous to the region than the tsunami caused by the impact (Schuiling et al., 2007).

4.4.3 Tsunamis in Lakes and Dams

Tsunami-like sea water disturbances have been also reported in lakes after strong earthquakes. For example, the strong, destructive *M*6.7 earthquake of February 18, 1911 that ruptured the area of Lake Ochrida in southern Jugoslavia, caused oscillation of the lake sea water with amplitude of about 0.5 m (AOA, 1912). According to Salamon (2004), the moderate *M*5.2 earthquake of February 11, 2004 in the Dead Sea caused a tsunami up to 1 m high and a rough sea in the northern Dead Sea basin. It is said that the large $M{\sim}7.0$ earthquake of January 1, 1837, that ruptured the area of Southern Lebanon and Northern Israel, caused waves that flooded the coast of Lake Tiberias, but it is not clear whether this happened before, during or after the earthquake (Ambraseys, 2009 and references therein).

An exceptional case is the one associated with the storage lake of Vajont Valley situated at the toe of Monte Toc, north Italy. The Vajont Dam was flooded over by more than 30 million m^3 of water. On October 9, 1963, a rockmass having a volume of approximately 250 million m^3 landslided into the dam, dislocated the water masses and caused a great flooding (Müller, 1964). The town of Longarone became a sea of mud that killed around 2000 people. Ward and Day (2010) extending the approach used for landslide-generated tsunamis and local runup to include overland flows and floods in a dam-break like situation, were able to model successfully the "tsunami" by considering that the landslide was not a debris avalanche, but rather a semi-coherent slump.

CHAPTER 5

What Do We Learn for the Source Characterization from Numerical Modeling?

5.1 INTRODUCTION

Extensive research has been performed in the field of tsunami wave numerical modeling in the last 30 years or so. The experience accumulated indicates that this kind of research, often combined with field observations, laboratory experiments, and benchmark testing and validation, is quite productive for a wide spectrum of tsunami studies. Tsunami simulations are a substantial help in developing a better understanding of the tsunami source type and generation mechanism through inversion techniques, in forecasting applications aiming to the good performance of tsunami early warning systems and in the calculation of tsunami wave propagation and inundation, which is extremely useful in studies of tsunami hazard, vulnerability and risk assessment. Of particular importance is the validation of reliable tsunami simulation codes through tests agreed upon by the tsunami community.

The relevant efforts made in the European and Mediterranean region are notable and find international recognition. The list is too long to mention all the codes developed for the numerical simulation of tsunamis and the relevant tests for validation. A good example is the code UBO-TSUFD developed at the University of Bologna, Italy. This solves the nonlinear, shallow-water equations in a Cartesian frame, with inclusion of bottom friction and exclusion of the Coriolis force, by means of a leapfrog finite-difference scheme on a staggered grid. It accounts for moving boundaries to compute sea inundation and withdrawal at the coast caused by several types of tsunami sources (Tinti and Tonini, 2013). Of particular interest is also the development of numerical modeling approaches for the simulation of tsunamis induced by rockslides, for example, in coastal fjord communities in Norway. Namely, Løvholt et al. (2015) demonstrated how they are able to model the short-frequency and high-amplitude tsunamis and to simulate their propagation and run-up in fjords for typical wave characteristics of rock-slide-induced waves.

Tsunamis in the European-Mediterranean Region. http://dx.doi.org/10.1016/B978-0-12-420224-5.00005-3

The progress achieved in the European and Mediterranean region with numerical simulation studies for the assessment of tsunami hazard, vulnerability, and risk is examined in Chapter 6. In Chapter 7 the contribution of tsunami simulations in the performance of early warning systems through forecasting applications is discussed. Therefore, the present chapter focuses on the contribution of tsunami numerical modeling in the characterization of the tsunami source and understanding of the tsunami generation mechanism through several cases of tsunamis induced by earthquakes, volcanic activities, and landslides.

5.2 CHARACTERIZATION OF TSUNAMI SOURCES FROM NUMERICAL MODELING INPUTS

Numerical simulation of tsunamis is a powerful tool in the effort to characterize the source and to discriminate between different types of sources, for example, seismic versus aseismic sources. The results, however, depend on many assumptions concerning the parameters of the source that are introduced in the calculations. From the methodological point of view, it is quite challenging to understand the possibilities and limitations of the various numerical models which are in use. The most productive way to go ahead with such an approach is to examine a number of modern tsunami events whose sources are controllable by relevant observational data. To this aim a set of characteristic seismic, volcanic, and landslide tsunami cases were selected for examination: (a) *seismic tsunamis*: December 28, 1908, Messina Straits; July 9, 1956, South Aegean; May 21, 2003, north Algeria; (b) *volcanic tsunamis*: LBA tsunami in Thera, South Aegean; (c) *landslide tsunamis*: February 7, 1963, west Corinth Gulf; December 30, 2002, Stromboli. Detailed information about these tsunami events, including relevant figures, can be found in various parts of Chapters 2–4.

5.2.1 Seismic Tsunamis

5.2.1.1 Messina Straits, December 28, 1908

This lethal tsunami was produced by a very strong, shallow disastrous earthquake that ruptured the Messina straits, South Italy, and measured a magnitude of *M*7.2. One of the most puzzling issues has been the precise location of the earthquake epicenter as well as the extent and geometry of the rupture zone. Figure 5.1 shows one of the very first, little-known suggestions for the 1908 tsunami source coming from the pioneering work of the Japanese seismologist F. Omori who immediately after the 1908 earthquake

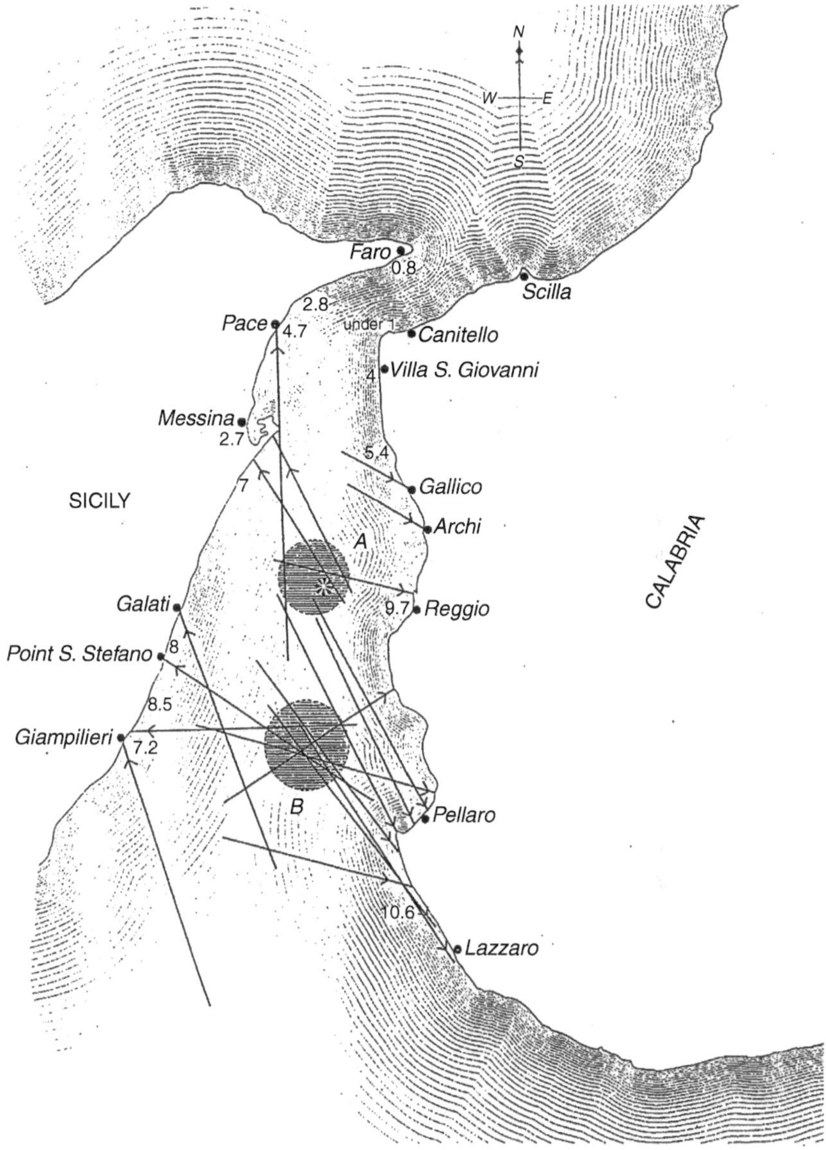

Figure 5.1 ***The two main sources of the December 28, 1908 tsunami and directions of the wave propagation towards coastal localities (after Omori, 1910).*** Figures at coasts indicate measured wave heights.

and tsunami traveled in the area and performed a detailed post-event field survey. The report of Omori (1910) was one of the very first published about the 1908 earthquake and its associated tsunami (see also the early reports by Baratta, 1910; Platania, 1909; Eredia, 1910). It can be observed that the tsunami source suggested by Omori (1910) does not differ significantly from the one proposed in modern studies (Billi et al., 2008, their Figure S1).

Extensive numerical simulation studies have been performed with the aim of understanding the source of the 1908 Messina tsunami. However, no conclusive results have been obtained so far given that it became impossible to definitely discriminate between a source involving only coseismic fault dislocation or seismically induced landslides or both, although many papers have been published for the seismic source modeling.

Seismic slip distribution models for understanding the 1908 earthquake source have been proposed by several authors based either on observations regarding coseismic elevations (Capuano et al., 1988; De Natale and Pingue, 1991; Amoruso et al., 2002, 2006; Pino et al., 2009) or on waveform modeling of historical seismograms (Pino et al., 2000). However, the various methodological approaches used have been debated (Amoruso et al., 2010; Pino et al., 2010).

As regards the investigation of the tsunami source mechanisms, Piatanesi et al. (1999), used two alternative tectonic source models and applied a scheme of finite-element simulations for the 1908 tsunami. Both sources were compatible with the polarity of the first impacting wave observed in some places along the coast of Sicily and Calabria. Nevertheless, although these authors took into account the actual bathymetry of the sea and topography of the coast, they found large discrepancies between the computed maximum water elevations and the observed maximum runups. They suggested that one reason for the discrepancies was that their model does not allow computation of the water runup. On the other hand, they supported that a heterogeneous rupture of the fault may be a reasonable source mechanism. As a matter of fact, an improvement was obtained by allowing a heterogeneous slip distribution on the fault plane (Tinti et al., 1999), which allow for a more realistic representation of the rupture process. However, the resulting tsunami was still too weak to explain the observations.

Recent hypotheses invoking submarine landslides opened a debate that is still ongoing with no conclusive and widely accepted interpretation (Billi et al., 2008, 2010; Gerardi et al., 2008; Argnani et al., 2009; Favali et al., 2009).

5.2.1.2 Amorgos, July 9, 1956

This is another destructive tsunami event observed after a large earthquake of *M*7.5 which ruptured the submarine trough between the islands of Amorgos, Santorini, Anafi, and Astypalea, south Aegean Sea (Figure 5.2).

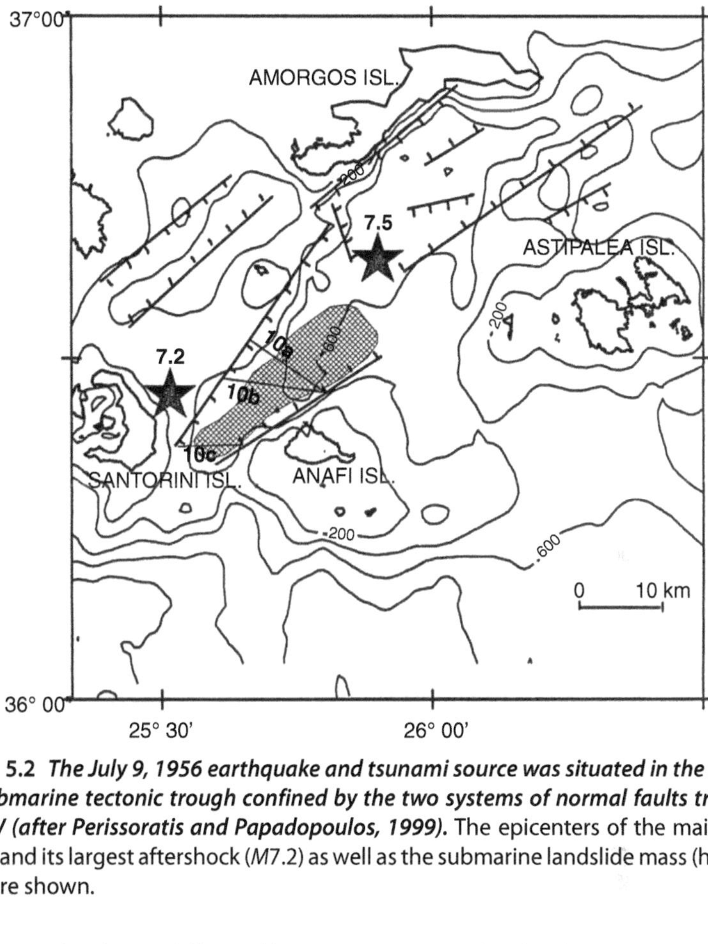

Figure 5.2 ***The July 9, 1956 earthquake and tsunami source was situated in the area of the submarine tectonic trough confined by the two systems of normal faults trending NE–SW (after Perissoratis and Papadopoulos, 1999).*** The epicenters of the mainshock (*M*7.5) and its largest aftershock (*M*7.2) as well as the submarine landslide mass (hatched area) are shown.

The main shock was followed by a strong *M*6.9 aftershock occurring 13 min later. From two near-field tide gage records (Figure 5.3a,b) in Lakki harbor, Leros Isl., and in Souda harbor, Crete Isl. (Figure 5.4a,b) Galanopoulos (1957a) and Ambraseys (1960) suggested that the tsunami was likely produced by submarine landslides triggered by the earthquake. A marine geophysical survey showed NE-SW trending normal faulting in the banks of the Amorgos basin, seafloor sediment instability, and a geologically very recent submarine landslide occupying the western side of the basin, with an area of 144 km^2 and volume of 3.6×10^6 m^3 (Perissoratis and Papadopoulos, 1999, Figure 5.2). The same authors calculated that the proximity of the landslide area to the earthquake epicenter implies seismic ground accelerations much higher than the minimum ones required to trigger landslide. Bell et al. (2012) estimated volume of 4.4×10^9 m^3. The landslide episode

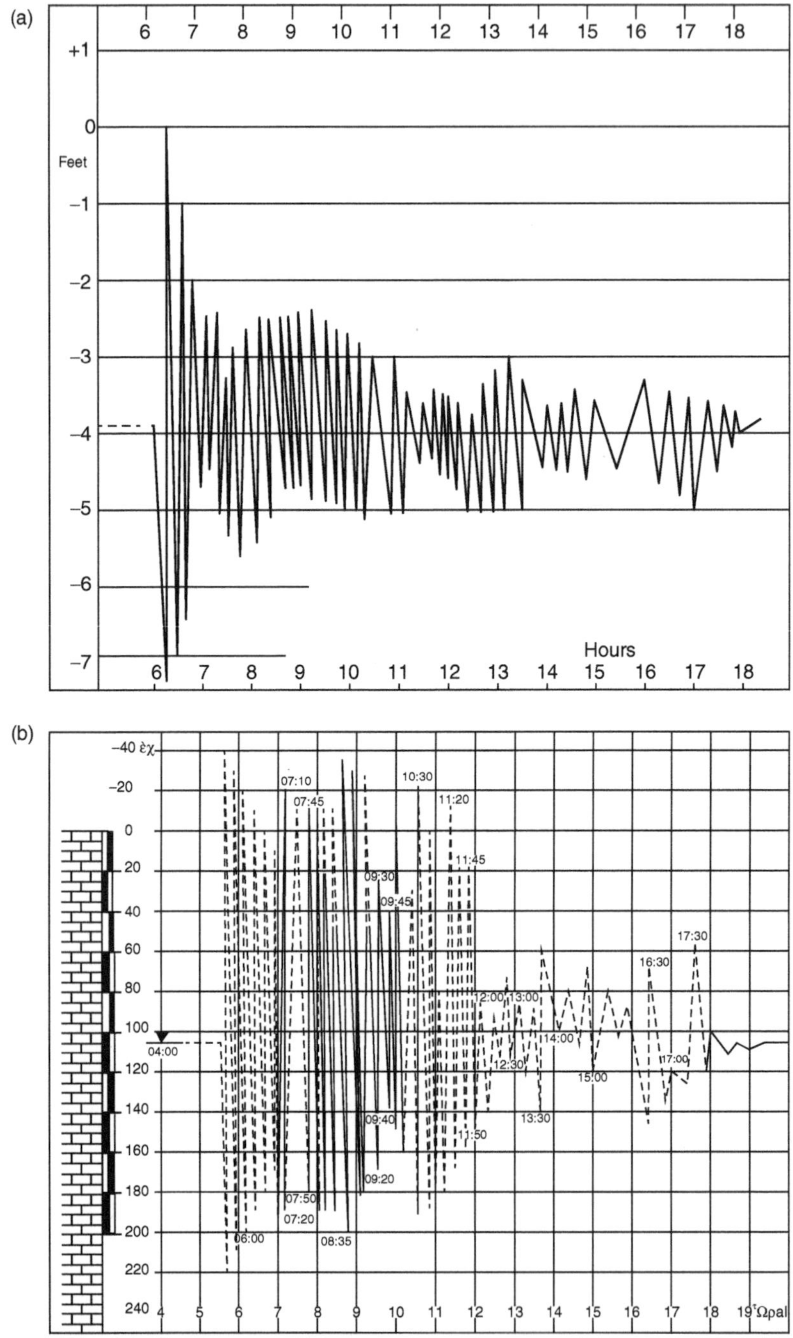

Figure 5.3 ***Tide gauges records of the July 9, 1956 tsunami in the harbors of Lakki, Leros Island (a) and of Souda, northwest Crete Island (b) (after Galanopoulos, 1957b).***

may have occurred in association with the large 1956 earthquake (Perissoratis and Papadopoulos, 1999) or with an earlier but post-Minoan strong earthquake or volcanic eruption (Bell et al., 2012) bearing potential for tsunami generation.

Numerical simulations based on the tectonic model showed a discrepancy by a factor of 3 to 10 between the nearly 30 m maximum heights initially reported (Galanopoulos 1957a,b; Ambraseys, 1960; see also Figure 5.4a) and the simulated ones ranging between 3 and 10 m at the source region (Yalciner et al., 1993; Okal et al., 2009). Field observations and interviews with eye-witnesses (Dominey-Howes, 1996; Papadopoulos et al., 2005a) showed that in Astypalaea Isl. the wave height may not have exceeded about 15 m. Therefore, part of the discrepancy could be explained by an overestimation of the initially reported wave heights. However, a significant part of the discrepancy remains unexplained. Consequently, an adequate reproduction of the near-field wave amplitudes requires not only coseismic seafloor fault displacement but also an additional tsunamigenic component such as massive coseismic landslide.

Beisel et al. (2009) simulated a landslide tsunami. The spectra of the resultant mareogram obtained in the proximity to the far-field Yafo tide gage station in Israel (Figure 5.5), contained harmonics with frequencies very close to those measured, thus favoring the landslide hypothesis. However, in the near-field domain this model produced only wave heights of less than 5 m. On the other hand, the simulation of a tsunami of tectonic nature resulted in amplitudes close to that recorded in Yafo, that is of about 28 cm at maximum (Figure 5.5), but did not contain significant spectral energy components with periods of ~15 min which appear in the tide gage record. Okal et al. (2009) tested also a series of landslides, but underestimated wave heights were again obtained in the near-field domain.

The introduction of heterogeneity in the seismic rupture process may affect substantially the tsunami run-up (Geist and Dmowska, 1999). Therefore, for the Amorgos tsunami, Papadopoulos et al. (2012b) examined the case of heterogeneous seismic slip on the fault by keeping the total seismic moment constant and considering two main sub-faults as shown in Figure 5.2. Following Bilek and Lay (1999), a depth-dependent shear modulus function was introduced in the earthquake generation area. Preliminary results from numerical simulation showed that the introduction of rupture complexity does not drastically change the near-field tsunami amplitudes. To explore further the landslide case as a triggering agent of the Amorgos tsunami, Papadopoulos et al. (2012b) simulated the tsunami by considering

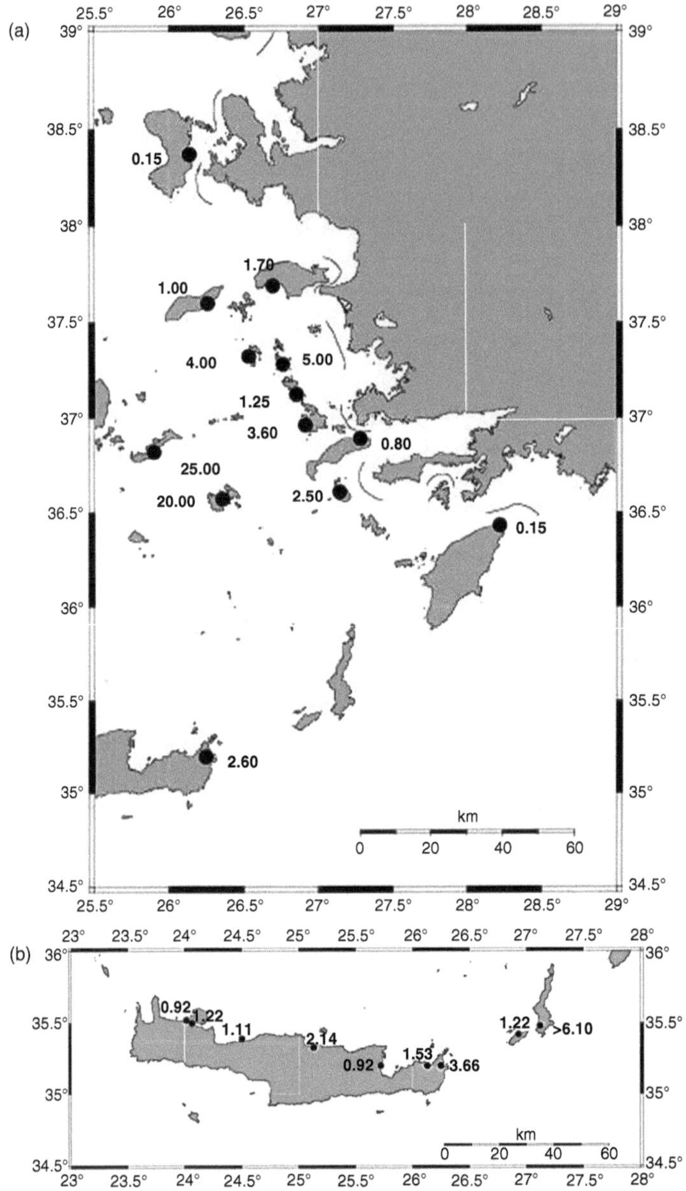

Figure 5.4 ***Wave heights (in meters) reported by Galanopoulos (1957b) for the July 9, 1956 tsunami to the east of the source.*** Maximum wave heights of 25 m and 20 m were reported in south Amorgos Island and in Astypalaea Island. Ambraseys (1960) reported wave heights of c. 30 m in south Amorgos Island. Wave heights (in meters) reported by Ambraseys (1960) for the July 9, 1956 tsunami in the islands of Crete, Kassos, and Karpathos, shown from east to west. In Souda harbor the wave height was 1.22 m. Similar values were given by Galanopoulos (1957b).

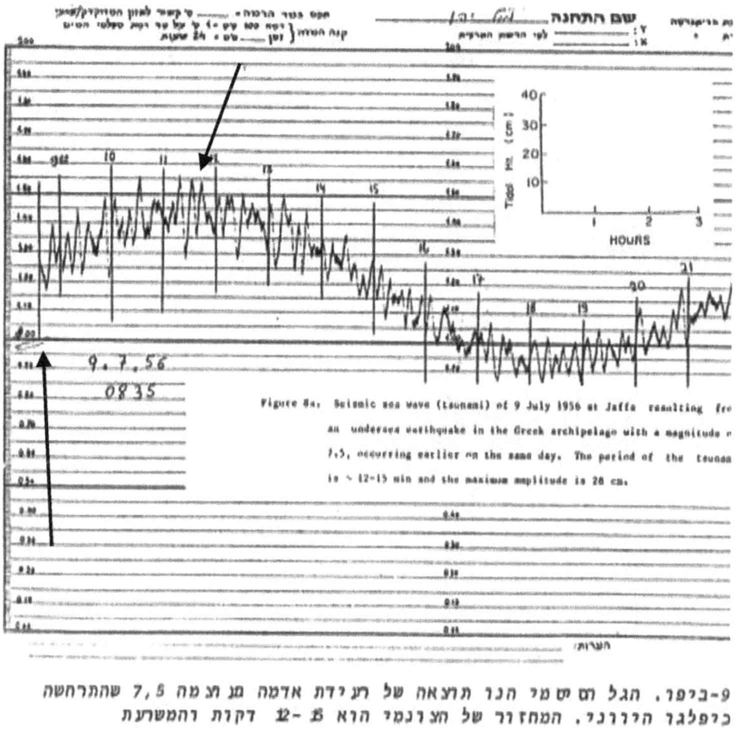

Figure 5.5 ***Tide gage record of the July 9, 1956 tsunami in the harbor of Yafo, Israel (after Goldsmith and Gilboa, 1985).*** The first arrow from left to right shows the earthquake origin time, while the next arrow shows the time of tsunami record.

the real submarine landslide identified by Perissoratis and Papadopoulos (1999) as well as four scenarios for landslides generated in the source area due to strong earth shaking. The tsunami produced by the real landslide underestimates the wave amplitudes observed while its polarity does not fit the wave polarity. On the contrary, at least one of the landslide scenarios produced a tsunami, which fits the observations. However, the landslide scenario underestimated drastically the wave amplitude obtained in the tide gage of Yafo in the far-field domain, which was well reproduced by the tectonic source model instead (Papadopoulos et al., 2012b).

5.2.1.3 Boumerdes-Zemmouri, May 21, 2003

The Boumerdes-Zemmouri, north Algeria, earthquake generated a small-to-moderate tsunami recorded at several localities around the coast of Balearic Islands, Spain. Studies by Yiga (2003) and Meghraoui et al. (2004) indicated that nonuniform slip occurred in the seismic fault and that the earthquake magnitude was larger than the M_w 6.8 assigned by the Harvard CMT solution. Wang and Liu (2005) suggested several fault plane solutions and developed a series of numerical simulations for the tsunami generation and propagation. Numerical results for tsunami wave heights were compared with the available tide gage measurements around Balearic Islands. Huge discrepancies were obtained between numerical results and the measurements and, therefore, an optimized model was developed which indicated that a larger moment magnitude (M_w7.2) is necessary to generate the tsunami wave heights observed at localities such as Eivissa (Ibiza) and Sant Antoni. However, the modeling performed by Alasset et al. (2006) concluded that the best fit between synthetic and real data is obtained for a thrust rupture of M_w6.9 comparable with the earthquake fault inferred from seismotectonic studies and located within 15 km offshore.

The misfit between the observed tsunami wave heights and the amplitudes recorded by tide gauges was also noted by Sahal et al. (2009) who performed a number of numerical simulation experiments to resolve the discrepancy. This effort, however, did not reach at positive results leaving the discrepancy rather unresolved. The authors concluded by emphasizing on (a) the importance of having accurate and precise data about the tsunami sources, (b) the need for networked tide gauge with high-resolution records and short sampling rates, and (c) the importance of conducting field studies immediately after a tsunami occurs.

5.2.2 Volcanic Tsunamis

5.2.2.1 Thera LBA Tsunami

The LBA tsunami caused by the giant eruption of Santorini (Thera) volcano, today documented by abundant geological signatures (see Chapter 2, e.g., Figure 2.14), was hypothetically caused either by a circular caldera collapse or by massive pyroclastic flow entering the sea water mainly from the southern side of the volcanic cone facing the north coast of Crete. Maximum tsunami wave heights calculated from numerical simulations based on the first hypothesis are of the order or may exceed 20 m, for example in the north coast of Crete, which is consistent with the archeological evidence and the tsunami sediment deposits found in several coastal localities of the eastern Mediterranean basin (Minoura et al., 2000, 2003; Bruins et al., 2008; De Martini et al., 2010; Novikova et al., 2011; Ozel et al., 2011; see review in Chapter 2). However, smaller wave heights were also calculated (Pareschi et al., 2006c). The pyroclastic flow hypothesis was tested by Novikova et al. (2011). The numerical simulation results are similar to those obtained for the caldera collapse mechanism as regards the maximum wave heights calculated.

Nevertheless, none of the two hypothetical mechanisms are absolutely realistic. In fact, the caldera collapse certainly may have produced a large tsunami but it is hardly understood how the wave propagated to several azimuths outside the caldera given that only one or two small gates to the open sea were existing in the north and the southwest sides of the caldera (Figure 5.6). The numerical simulations performed so far do not take into

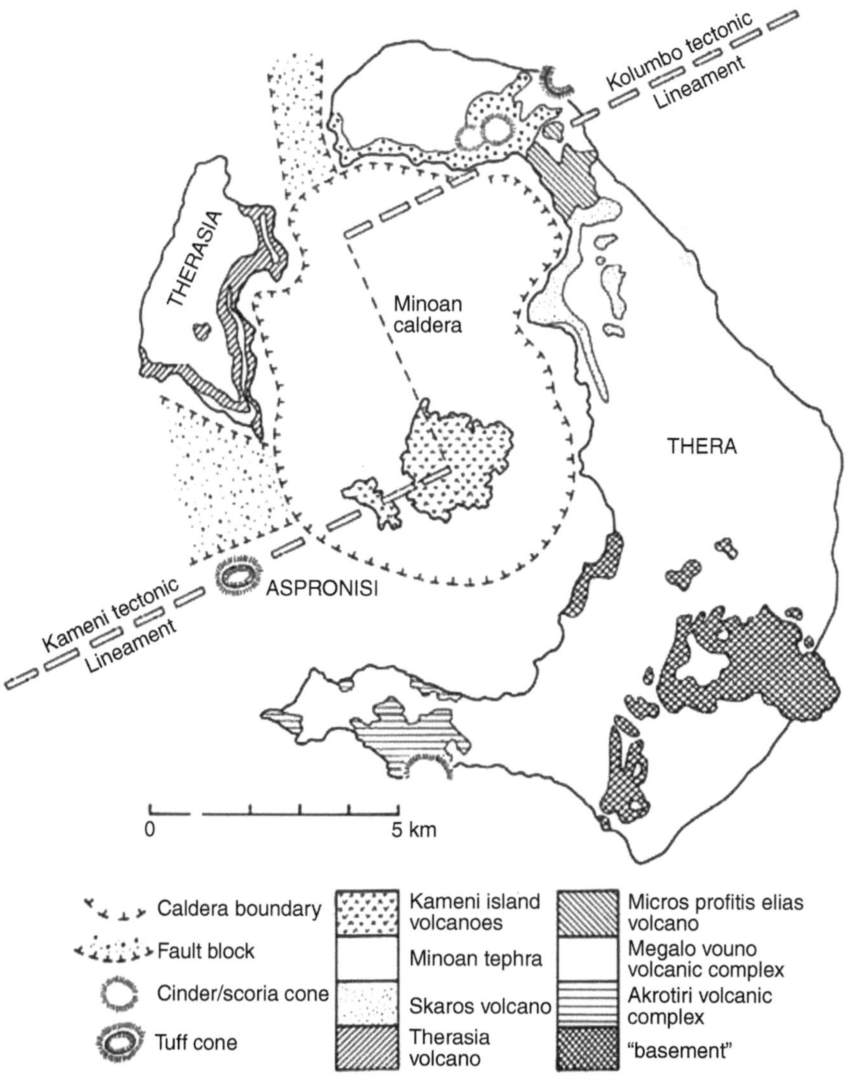

Figure 5.6 ***Map identifying the generalized geology of Thera (after McCoy and Heiken, 2000).***

account this important physiographic feature, considering only a simple, open geometry for the tsunami source.

On the other hand, the results reached at by Novikova et al. (2011) showed very strong energy directivity as well as strong dependency of the tsunami height on the azimuth of pyroclastic flow penetration in the sea (Figures 5.7 and 5.8). These results may interpret the tsunami sediment deposits found in north Crete (Minoura et al., 2000; McCoy and Papadopoulos, 2001; Bruins et al., 2008) but leave inexplicable how tsunami sediments were deposited in coastal sites situated at large distances and at very different azimuths with respect to the suggested direction of pyroclastic flow penetration into the sea water (see review in Chapter 2). This puzzling issue was recently noted by Papadopoulos et al. (2014a) and Periáñez and Abril (2014).

Several years ago some authors suggested that the LBA tsunami may have been triggered by large earthquake(s) that associated the great eruption of Thera (Antonopoulos, 1992a; Pararas-Carayannis, 1992). However, no seismotectonic evidence was given to support such a hypothesis. Periáñez and Abril (2014) re-considered the earthquake scenario and suggested that earthquake(s) associated with the LBA eruption or even the caldera collapse itself may have triggered either massive deep-sea turbidites offshore Nile Delta (Ducassou et al., 2009, 2013) or the so-called Augias megaturbidite recognized in the Ionian abyssal Plain (Cita et al., 1996; see review in Chapter 2) which turned to act as tsunami triggering agents. Numerical simulations showed that the scenario of Nile Delta turbidites is the most favored to explain tsunami sediment deposits found in the eastern Mediterranean basin outside the Aegean Sea domain (Periáñez and Abril, 2014). However, this possibility still remains hypothetical as more field data are needed to support it.

5.2.3 Aseismic Tsunamis

5.2.3.1 Corinth Rift, 1963

The landslide of deltaic deposits of low consolidation, which occurred in the south coast of the western Corinth Gulf on AD 7 February 1963 (Figure 5.9, see also Figure 3.8 in Chapter 3), is a characteristic example of purely aseismic landslide producing local but still powerful, damaging tsunami with a maximum wave height up to about 6 m. A post-event field survey provided a good dataset regarding the observed impact of the tsunami in both coasts of the gulf as well as precise estimates of the area, volume, and other physical parameters of the failed mass (Galanopoulos

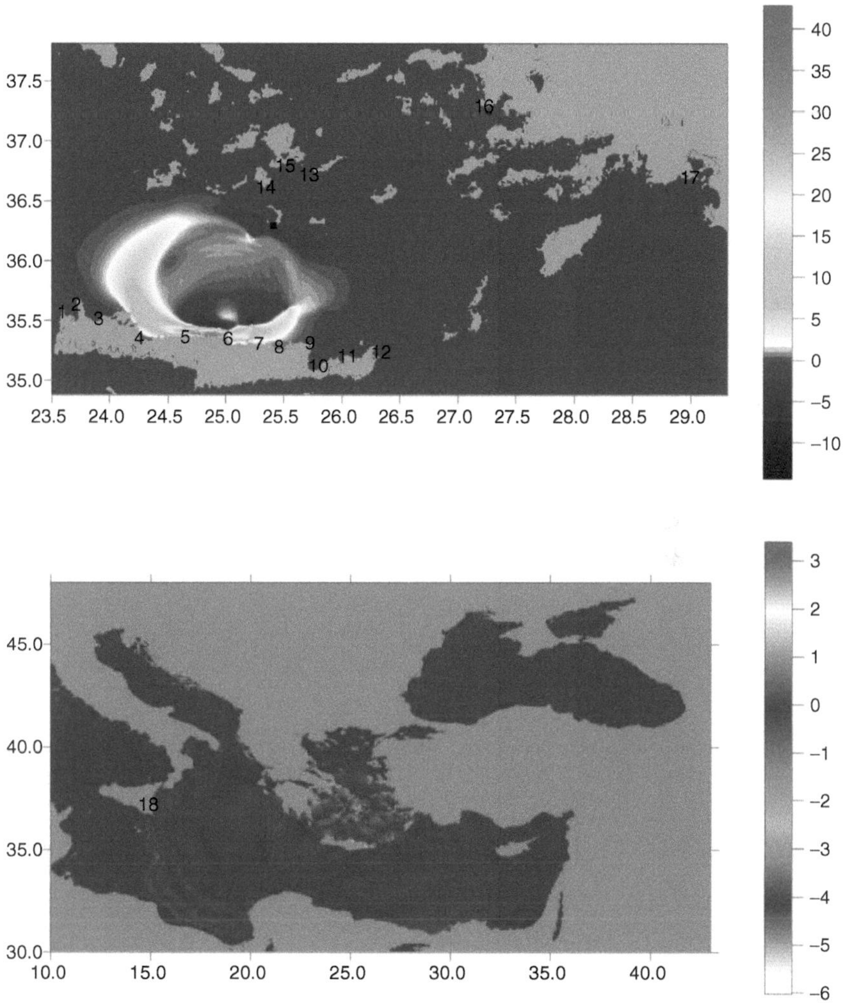

Figure 5.7 ***Snapshots of Thera simulated tsunami wave about 13 min after its generation (after Novikova et al., 2011).*** The tsunami was produced by thick (~55 m) pyroclastic flow entering the sea from the south of Thera with direction 145° measured from north counter-clockwise. Colored scales are in m. The location of pyroclastic flow at Thera shore is marked by the solid square. Wave profiles from synthetic tide gage records are shown in Figure 5.8. Twelve tide gauge stations are placed offshore north Crete coast above sea depth of ~20 m (marked from 1 to 12). Another three locations are situated in Cyclades islands (marked from 13 to 15), and another two are situated in western Turkey, namely in Didim (16) and Fethiye (17). Number 18 (in the lower panel) shows the coastal locality of Augusta in eastern Sicily (for more details see in Novikova et al., 2011).

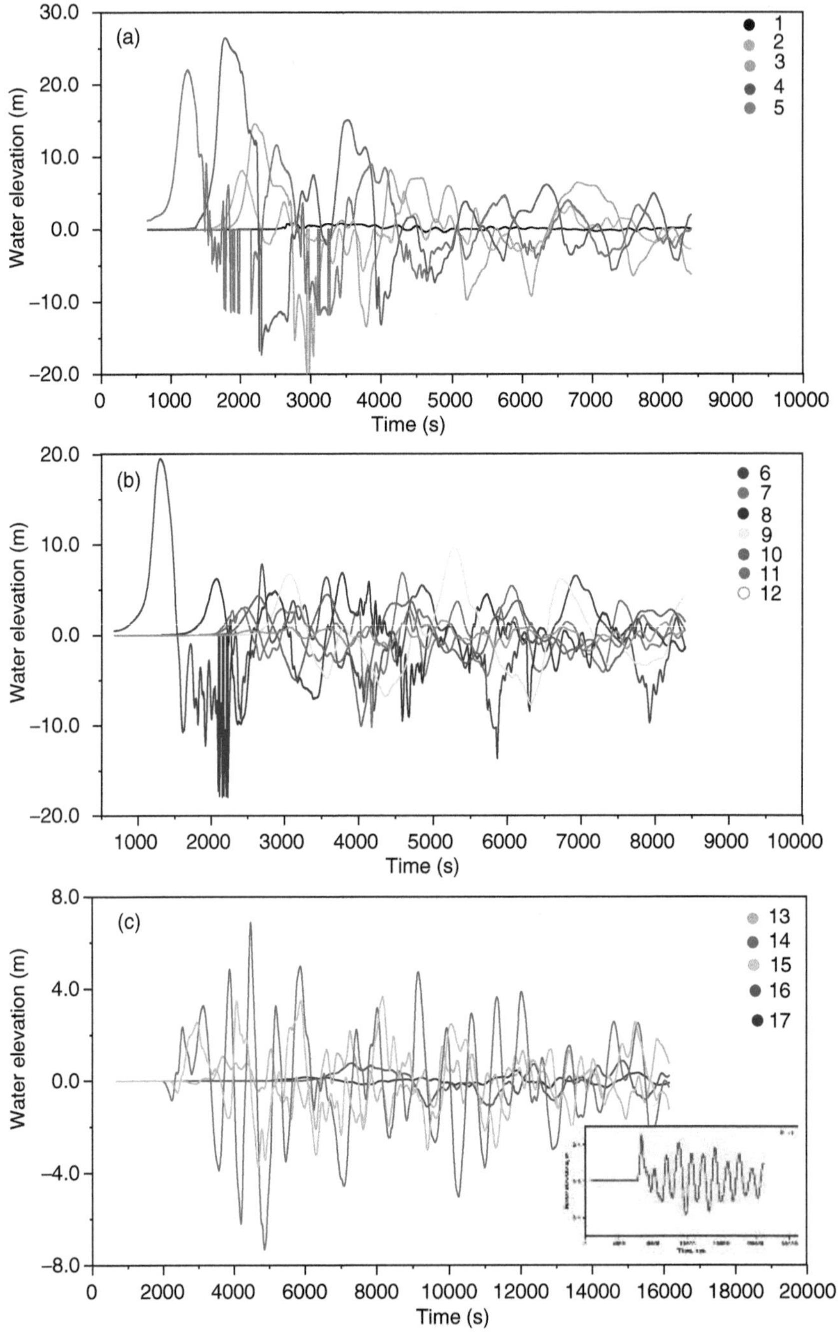

Figure 5.8 ***Synthetic wave profiles from the gauge records numbered from 1 to 18 (Figure 5.7) (after Novikova et al., 2011). Numbers from 1 to 17 in subparts a, b, and c correspond to tide gauges numbered in Figure 5.7. Inset in subpart c shows wave profile at synthetic tide gauge station numbered 18 in eastern Sicily.***

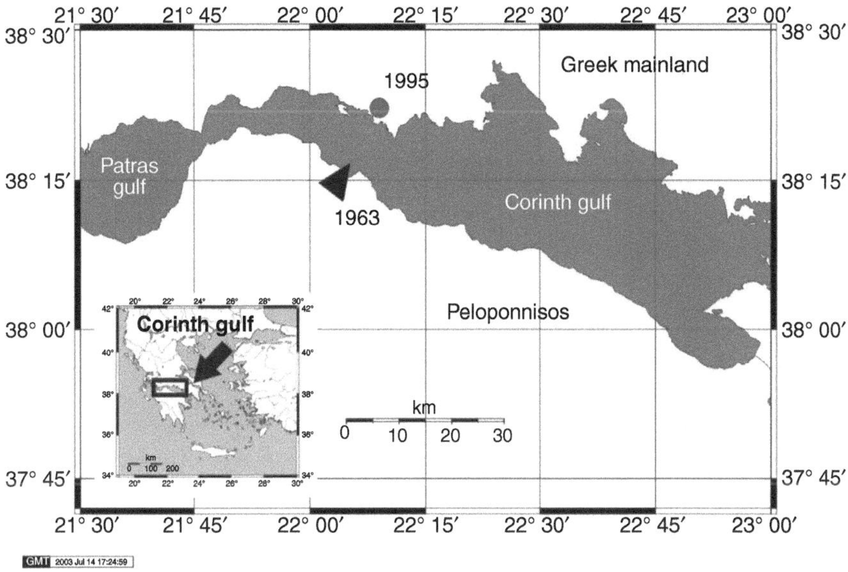

Figure 5.9 ***The western Corinth Gulf, Central Greece, where the tsunami of February 7, 1963 tsunami was observed as a result of an aseismic coastal landslide shown by a solid triangle.*** On June 15, 1995 a smaller local tsunami was generated by a *M*6.2 earthquake (epicenter shown by red solid circle).

et al., 1964). Numerical simulations of the tsunami considering either a rigid-body model for the landslide (Koutitas and Papadopoulos, 1998) or a layered sediment structure set in motion by an initial external dynamic action (Papadopoulos et al., 2007c) reproduced well-enough the observed runup data in the field. However, the second approach permits to describe in detail the formation of several wave groups and particular tsunami characteristics that are strongly dependent on the landslide model.

5.2.3.2 Stromboli, December 30, 2002

Details on this tsunami that was produced by a volcanic landslide in Stromboli volcano, Aeolian Arc, can be found in Chapter 2. Maximum runup of about 9 m was measured (Figure 5.10). This case is an excellent example of landslide tsunami reproduced by numerical simulations. In fact, results of numerical modeling fit very well both the observed phenomena (Maramai et al., 2005b) and the experimental data (La Rocca et al., 2004; Tinti et al., 2006b) thanks to that the source process, that is the sequence of two landslides, is very well constrained.

Figure 5.10 ***In Stromboli a maximum runup of ca. 9 m was measured during a post-event field survey performed jointly by scientists of the University of Bologna (Italy), the Tohoku University (Japan), and the National Observatory of Athens (Greece) (photo by G.A. Papadopoulos).***

5.3 SOME CONCLUSIONS

Results regarding the 1908 Messina Straits, 1956 Amorgos and 2003 Algeria seismic tsunamis disfavor any conclusive remark about the source mechanism which clearly indicates the complexity of the source and the need for more research. A critical next step would be considering the superposition of two waves produced by a combined source consisting from a coseismic displacement coupled with a landslide.

On the other hand, the cases of 1963 in Corinth Gulf and of 2002 in Stromboli leave no doubt that the good knowledge of the tsunami causative landslide source mechanism is of substantial importance for the effective reproduction of the tsunami by applying numerical simulation techniques.

CHAPTER 6

Hazard, Vulnerability, and Risk Assessment

6.1 INTRODUCTION

Coastal communities are exposed to hazards due to sea level changes while the associated risks increase as the communities become more and more developed. In the case of tsunamis, hazard, vulnerability, and risk, modeling and assessments are essential to develop effective tsunami risk mitigation and management systems that may offer appropriate guidelines for emergency planners and decision makers.

In the European and Mediterranean region (EM), only sporadic tsunami studies and publications were undertaken since the 1950s. Coordinated tsunami research was put forward since 1992 with several projects funded by the European Commission, such as the GITEC (1992–1994), GITEC-TWO (1996–1998), TRANSFER (2006–2009), SCHEMA (2007–2010), and ASTARTE (2013–) projects. However, only after the big Indian Ocean tsunami of December 26, 2004 a systematic effort has been made to develop actions aiming to reduce tsunami risk. The main initiatives are the North-East Atlantic and Mediterranean Tsunami Warning System (NEAMTWS) operating under the coordination of the country members affiliated with the IOC/UNESCO and the Tsunami Programme developed by the Joint Research Center (EU) and is supported by the European Commission (EC) (Chapter 7). Collateral actions include education and preparedness projects, some of them coordinated by the NEAMTIC, the Tsunami Information Center in the North-East Atlantic and Mediterranean (NEAM) region, as well as by parallel national actions.

This chapter focuses on issues related to the tsunami hazard, vulnerability, and risk assessment, particularly as regards in the NEAM region. However, the international experience obtained beyond NEAM is not ignored since it is quite useful for better understanding the current status in this field of research in the NEAM region.

It is noteworthy that a puzzling issue is the very different ways that terms such as hazard, exposure, vulnerability, risk, impact, damage, resilience, and similar are in use by different authors or listed in glossaries prepared

Tsunamis in the European-Mediterranean Region. http://dx.doi.org/10.1016/B978-0-12-420224-5.00006-5

by several organizations. Therefore, the next section is devoted to a critical discussion of terminologies.

6.2 GLOSSARY AND TERMINOLOGY

In the science of natural hazards, a variety of terms are in use to express the level of exposure to this or that type of hazard as well as the level of impact that could be produced. Although there is a general consensus on the need to discriminate between hazard and risk, often confusion prevails as regards these terms as well as for other relevant terms in use, such as exposure, vulnerability, resilience, damage, and intensity. Therefore, it is useful to look at definitions suggested for such terms and indicate what are the most appropriate to adopt.

A comprehensive report relevant to the issue of guidelines and definitions related to natural hazards and risks is the "Commission Staff Working Paper" produced by the EC (2010), hereafter called EC-Working Paper (2010) for reasons of brevity. For the purpose of these guidelines, international standards developed by the International Organization for Standardization, in particular ISO 31000, ISO 31010, and the corresponding ISO guide 73 for terminology were used, combined along with the more targeted UNISDR terminology on disaster risk reduction.

6.2.1 Tsunami Hazard

According to the EC-Working Paper (2010) ***natural hazard*** *is a natural process or phenomenon that may cause loss of life, injury, or other health impacts, property damage, loss of livelihoods and services, social and economic disruption, or environmental damage*. The term is used to describe actual hazard events as well as the latent hazard conditions that may give rise to future events. Natural hazard events can be characterized by their magnitude or intensity, speed of onset, duration, and area of extent (UNISDR, 2009).

In the tsunami glossary coordinated and published by IOC (2006, 2013) ***tsunami hazard*** *is the probability that a tsunami of a particular size will strike a particular section of coast*, while the ***tsunami hazard assessment*** *includes an identification of populations and assets at risk and definition of the level of that risk*. However, since the hazard assessment, regardless of whether it concerns tsunami, earthquakes, or other natural phenomena, could be performed either on a qualitative way or quantitatively from probabilistic models or other approaches, for example, deterministic ones, my view is that the definition of the term hazard should not rely on the method(s) in use (e.g., probabilistic) for the hazard assessment but on the fact that hazard is an expression only of a phenomenon that potentially may have some impact.

On the basis of the previous analysis, I do prefer to define ***tsunami hazard*** *as a qualitative or quantitative expression of the possibility that potentially tsunami impact may take place in a particular coastal segment, area, or region within a given time window.* The assessment of the tsunami hazard level can be based on descriptive, statistical, probabilistic (stochastic), or deterministic approaches.

6.2.2 Tsunami Exposure and Vulnerability

In relation to the previous considerations and by taking into account definitions of terms proposed by UNISDR (2009) the term ***exposure to tsunami hazard*** *may express people, property, systems, or other elements present in tsunami hazard zones that are thereby subject to potential losses.* As a consequence, *the characteristics and circumstances of a community, system, or asset that make it susceptible to the damaging effects of tsunami hazard is termed* ***tsunami vulnerability.*** To make it fully understandable two examples are given next.

Vulnerability example 1, Buildings: Suppose that two houses, one of reinforced concrete frame and another rural, are lying at exactly the same point of a coastal zone and are attacked by a powerful tsunami. In this case all natural conditions are the same for both houses but their characteristics, that is their physical features, are different. It becomes obvious that the houses are not characterized by the same degree of vulnerability; the one of reinforced concrete frame being more tsunami-resistant than the rural one. As an alternative, let us assume that the second house is lying significantly further inland with respect to the first one. The characteristics of both buildings remain unchanged. What changes is the degree of exposure which is much higher at a distance of only 5 m from a fixed point of the seashore and much less at a distance of 1000 m inland from the same seashore point. In other words, the circumstances under which the two buildings are exposed are quite different and increase the relative vulnerability of the first building with respect to that of the second.

Vulnerability example 2, Humans: Suppose that a father and his daughter are standing in a particular point of a coastal segment at the time that a powerful tsunami attacks the coast. As soon as they realize the tsunami threat they need to counteract. A realistic option is to run to higher ground. In this realization, the daughter becomes more vulnerable compared to her father. In fact because of her natural features, she is not able to understand the danger immediately, as the father does, and she is unable to run as quickly as her father. However, one may argue that the father is more vulnerable by considering that his daughter got school lessons about the correct personal behavior regarding tsunamis but the father did not. From this point of view, education plays an important role. The father's physical vulnerability is still less than that of his daughter but he is more exposed to the tsunami hazard because he is uneducated.

6.2.3 Tsunami Impact, Damage, Intensity, and Resilience

The consequence of the previously given definition of the term vulnerability is that keeping exactly the same features for the potentially destructive natural phenomenon, here tsunami is meant, two or more assets of the same type (human beings, houses, or other engineered structures, vessels, etc.) have different overall vulnerability, not only because of their different physical features but also due to the different degree of exposure which is controlled by a variety of other factors.

In the present discussion, according to the EC-Working Paper (2010) ***human impacts*** *due to natural hazards are defined as the quantitative measurement of the number of deaths, severely injured or ill people and permanently displaced people.* On the other hand, ***economic impacts*** can be quantified in terms of immediate or long-term sum of costs, while ***political, social, and environmental impacts*** are usually difficult to quantify and, therefore, they are expressed qualitatively or semiquantitatively.

A relevant term is ***tsunami damage,*** which is the *loss or harm caused by a destructive tsunami* (IOC, 2013), while ***tsunami intensity*** *is the degree of tsunami impact on the anthropogenic and the natural environment estimated in a certain scale* (see review and introduction of a new 12-point intensity scale in Papadopoulos and Imamura, 2001). A more general definition of tsunami intensity considers it as a measure of strength, force, or energy (IOC, 2013) but such terms are respective rather to physical hydrodynamic tsunami features than to the impact itself.

Another important concept is that of ***resilience***, which according to the IPCC (Intergovernmental Panel on Climate Change, e.g., Denton et al., 2014), is *the capacity of social, economic, and environmental systems to cope with a hazardous event or trend or disturbance, responding or reorganizing in ways that maintain their essential function, identity, and structure, while also maintaining the capacity for adaptation, learning, and transformation*. The term resilience was seen from a similar point of view by the UNISDR (2009) and adopted by the EC-Working Paper (2010): ***resilience*** *is the ability of a system, community or society exposed to hazards to resist, absorb, accommodate to and recover from the* *effects of a hazard in a timely and efficient manner, including through the preservation and restoration of its essential basic structures and functions.*

A remarkable approach was presented by Flood and Schechtman (2012) who examined the value of resilience as a concept in planning and policy in the USA and Ireland associated with climate change. Three alternative approaches were put forward: engineering, ecological, and psychological resilience, while the resilience of each of these systems is sometimes at odds with each other. In that study the authors argue that "becoming more resilient" requires strengthening all three systems in a reinforcing manner rather than championing one to the detriment of the others.

As regards ***tsunamis***, a more compact definition could be read as: *tsunami resilience is the ability to withstand to, and recover from, the impact caused by tsunami action.*

6.2.4 Tsunami Risk

To discriminate between hazard and risk, the EC-Working Paper (2010) adopted that according to ISO 31010 ***risk*** *is a combination of the consequences of an event (hazard) and the associated likelihood/probability of its occurrence.* Mathematically this may be expressed as a convolution of two terms:

$$\text{Risk} = (\text{hazard impact}) \star (\text{probability of occurrence}) \tag{6.1}$$

Such a perception of risk is not new. For example, a similar expression was introduced by Smith (1992). However, with the aim of getting quantitative results reflecting a wide range of the potential impact, we may consider that risk could be expressed as the convolution of three main factors, thus taking the form

$$\text{Risk} = \text{hazard} \star \text{vulnerability} \star \text{value} \tag{6.2}$$

In this sense, we see that vulnerability is a critical quantifiable factor in the overall risk assessment. Then, risk is expressed in a descriptive, qualitative, or quantitative way depending on how the three factors composing it are expressed. If the term hazard incorporates the percentage probability and at the same time vulnerability and value are quantified, then we may define ***tsunami risk*** *as a quantitative description of the expected impact of tsunami action in a particular coastal segment, area, or region within a given time window.* One may realize that in this sense risk is dependent on the level of the physical process or phenomenon, that is on tsunami hazard, on the vulnerability of the several elements or assets exposed to the tsunami action, as well as on the economic (or other) value(s) exposed to hazard.

From the previous analysis it comes out that the increase of resilience results in the decrease of risk and vice versa. Then, for reasons of simplicity we may mathematically express it as

$$\text{Resilience} = \frac{1}{\text{risk}} \tag{6.3}$$

Combining (6.2) and (6.3) we see that keeping constant the level of hazard, then both risk and resilience are directly dependent on vulnerability and value, that is on a variety of social, economic, and environmental parameters.

6.3 TSUNAMI HAZARD ASSESSMENT IN THE EM REGION

The issue of tsunami hazard in the EM region has been seen in different ways by different authors. The several approaches used for the assessment of the tsunami hazard can be classified in three gross categories: descriptive or qualitative evaluations, statistical, and probabilistic approaches, scenario-based assessments.

6.3.1 Qualitative Evaluations

Many authors focused their attention on the descriptive analysis of the various types of data that may offer a good basis for the qualitative tsunami hazard evaluation. Very often the data come from historical documentary sources. The contribution of such data sets, usually presented in the form of parametric catalogs published in papers and books or organized in relevant web sites, has been extensively discussed in Chapter 2. In a later section, it is shown how parametric catalog data are utilized for the statistical and probabilistic tsunami hazard assessment in the EM region.

Of importance are also the geological and archeological signatures of past tsunamis, which may offer a basis to better estimate the repeat times of the large tsunami events. An example of a relevant discussion is the paper by Dawson et al. (2004), although it does not offer an exhaustive description of the several cases studied in the EM region. Of particular interest for the evaluation of the tsunami generation potential is the knowledge of the earthquake source properties and of the deformation mode (Tinti, 1990; Papazachos and Dimitriu, 1991; Pondrelli et al., 1998; Yolsal et al., 2007; Yolsal and Taymaz, 2010; Ebeling et al., 2012; Yolsal-Çevikbilen and Taymaz, 2012; Papadopoulos et al., 2014a). The integration of geologic fault data into tsunami hazard studies is an emerging topic of particular interest (Basili et al., 2013). The relative sea level changes play an important role in calibrating results obtained for the expected tsunami runup. Sea level changes could be due to long-term tectonic activity (e.g., in NW Greece, Vött, 2007) or eustatic sea level changes (Flemming and Webb, 1986), to storm surges (Dawson et al., 2004) or abrupt coseismic uplift or subsidence (e.g., in Phalasarna, NW Crete, see details and references in Chapter 2). Another approach is to combine satellite images with DEM (Digital Elevation Model) for the purpose to indicate coastal areas that may be prone to tsunami flooding (Theilen-Willige, 2008). A GIS method for the assessment of rock slide tsunami hazard in all Norwegian lakes and reservoirs was published by Romstad et al. (2009).

6.3.2 Statistical Recurrence From Paleotsunamis

6.3.2.1 Eastern Mediterranean

Geological signatures of tsunamis either inland or offshore is a valuable tool for the identification of sizable paleotsunami events, thus extending the historical tsunami record in the geological past. The consequences for the calculation of the tsunami recurrence period is obvious. In Chapter 2, an extensive review of geological paleotsunami studies performed in the EM region is presented. However, there are several problems that someone should be aware of when using geological paleotsunami results. Dating is a common problem. A characteristic example concerns the 20–25 m thick megaturbidite, known in the literature as Homogenite/Augias that was identified by geophysical surveys and sediment cores collected from the Ionian Sea bottom. In Chapter 2, one may find extensive discussion on the various relevant references, which showed that the Homogenite/Augias was triggered by the large tsunami produced by the LBA (late seventeenth century BC) Santorini caldera collapse. However, more recent results supported that the homogenite was rather caused by the AD 365 Cretan mega-earthquake/tsunami (Polonia et al., 2013) that is about 2000 years later than initially thought.

In the eastern Hellenic Arc there is a good case for the correlation of historical reporting with geological tsunami evidence. In fact, from the beginning of the fourteenth century up to now, that is in a time segment for which the catalog of strong tsunamis could be considered as complete, four strong tsunamis were reported to hit the area of Rhodes Island, Greece: AD 1303, 1481, 1609, and 1741 (Chapter 2). In Dalaman, SW Turkey, at the opposite site of Rhodes, a series of three sea sand layers were discovered and correlated with the historically documented tsunamis of 1303,

1481, and 1741 by geological methods (Papadopoulos et al., 2005b, 2012a) and biomarkers (Alpar et al., 2012). Only the historical tsunami of 1609 is missing from the stratigraphy record. Two alternatives may explain this: the 1609 tsunami either inundated the Dalaman coastal zone but did not leave behind tsunami deposits in the trenching site or failed to inundate at all. Consequently, the geological record of tsunamis appears extremely sensitive to several possible factors. In Fethiye, located only about 20 km to the east of Dalaman, we were able to identify only LBA tsunami deposits (Minoura et al., 2000). In Kamari residential zone, east coast of Thera (Santorini) island, South Aegean Sea, Dominey-Howes et al. (2000) investigated several trenches and no tsunami traces were found. On the contrary, at a distance of no more than about 500 m from one of the investigated trenches, our team was able to identify tsunami deposits very possibly attributable to the large tsunami of September 30, 1650 (old style) associated with an eruption of the submarine Kolumbos volcanic center (Papadopoulos, 2009). The sensitivity of the tsunami sediment deposition is strongly controlled by factors like coastal configuration, local topography, and ground erosion.

The earthquake sedimentary record in the seabed is of great importance for the identification not only of paleotsunamis but also of paleoearthquakes (Drab et al., 2012). Turbidite record collected by gravity cores in the Calabrian Arc subduction complex, Ionian Sea, indicated that the turbidite emplacement was triggered by three historically reported earthquakes in the area, that is, AD 1908 in Messina Straits, AD 1693 in NE Sicily, and AD 1169 which also hit Messina (Polonia et al., 2012). Along the coast of eastern Sicily, through a multitheme approach (Chapter 2) De Martini et al. (2010) were able to identify a record of 6–7 tsunamis in about the last 4 ka. In the historical period the tsunami sediment layers were correlated with the historically known tsunamis of AD 1693 local one, and the distant events of AD 365 in Crete and of LBA in Thera. In the Augusta Harbor, east Sicily, Smedile et al. (2011) worked on a 6.7 m long, fine sediment core sampled 2.3 km offshore at a water depth of 72 m (Chapter 2) and concluded that 12 anomalous layers could be attributed to prehistorical and historical tsunamis. The historical tsunamis, which correlated with the geological evidence include again the local tsunamis of AD 1908, 1693, and 1169 and the basin-wide ones of AD 365 Crete and the LBA of Thera.

The lesson learned from the cases of eastern Hellenic Arc and east Sicily is that the presence in a particular area of a series of past tsunami events, identified by both historical and geological evidence and covering a wide time window, certainly facilitates the calculation of the tsunami recurrence period.

On the other hand, however, often paleotsunamis and historical tsunamis are uncorrelated. From this point of view, the Ionian Sea on the side of Greece is a very characteristic case. In fact, in several papers it has been supported that a series of supposedly strong or even megapaleotsunami events, covering a time span extending from late Holocene up to the historical period, were geologically identified in various coastal and lake environments: Ambrakian Gulf (NW Greece) occurring around 2870–2350, 1000, and 300 cal. BC as well as around 840 cal. AD (Vött, et al., 2007); north side of Lefkada Island (central Ionian Sea, Greece) occurring between 395 cal. BC and 247 cal. BC (Vött, et al., 2007); Lake Voulkaria (Akarnania, NW Greece) occurring in the sixth millennium BC or shortly afterward, in around 1000 cal. BC as well as around AD 365 (Vött, et al., 2009); Palairos–Pogonia Bay (Akarnania, NW Greece) occurring in the sixth, fifth, fourth, and third millennium BC as well as in the fourth/third century BC (Vött, et al., 2011a); and harbor of ancient Krane (Kutavos Bay, Cefalonia Island, central Ionian Sea, Greece) occurring in historical times (Hadler et al., 2011).

The paleotsunami results obtained for the Greek side of the central Ionian Sea by the previously mentioned authors are quite challenging for a number of reasons. The first is the frequency of large tsunami occurrence. For example, in a regional scale in Akarnania the recurrence interval for large events was calculated to ca. 500–1000 years (Vött, et al., 2011). The overall geological estimate of tsunami recurrence from the sixth millennium up to recent historical times in the central Ionian Sea indicates a very high rate with respect to the tsunami record from documentary sources. In fact, in the historical record the generation of only a few minor tsunami events has been reported there (Papadopoulos and Chalkis, 1984), although the level of seismicity is very high (Papadopoulos and Kijko, 1991). This fact was already noted in Chapter 4 and attributed to that in areas of high seismicity which are dominated by active strike-slip faulting, such as in the central Ionian Sea as well as in the North Aegean Sea trough, the tsunami activity is significantly low.

In view of these considerations, the paleotsunami record in the Greek coastal zones of the central Ionian Sea is in strong contrast to the historical record. The contrast is not eliminated by the possible propagation toward the central Ionian Sea of the large AD 365 tsunami, which had its source further to the south in the area of western Crete (Figure 2.2). Yet, historically this tsunami has not been reported to hit the central Ionian Sea, although according to a local legend it may have reached up to the Dalmatian coast (Guidoboni et al., 1994).

In a similar seismotectonic environment of strike-slip faulting in the NE Aegean Sea, tsunamigenic sedimentary layers described there remain orphan because they are not correlated with historically documented tsunamis (Mathes-Schmidt et al., 2013), thus putting again in doubt if the criteria used for the discrimination of tsunami sediment layers from other stratigraphic sequences are correct.

6.3.2.2 The Olympia Tsunami Hypothesis

Of relevance to the investigation of paleotsunamis in the Ionian Sea is the "Olympia tsunami hypothesis" introduced by Vött et al. (2011a) with the aim *to clarify the mystery of the rapid burial of Olympia under 4–6 m of sediments after the sixth century AD and subsequent erosion of the Kladeos River by 8–10 m down to the ancient flow level* (for geographical references of this section, see Figure 6.1). These authors performed sedimentological, geophysical, geochemical, and microfaunal analyses and concluded that the Kladeos River valley and the classic archeological field of Olympia, NW Peloponnese, experienced at least four distinct phases of catastrophic high-energy flood events attributed to multiple tsunami impact. Therefore, they set up their hypothesis by suggesting that the shallow saddles of the Flokas–Platanos ridge, downstream Kladeos, were repeatedly overflowed by tsunami waters and the cult site of Olympia was rather destroyed by tsunami than by fluvial processes related to the Kladeos River. Since a charcoal fragment from a fining upward sequence out of sand and gravel deposited under high-energy conditions at the ancient Olympia environs was ^{14}C AMS radiocarbon dated to cal. year AD 585–647, at 2σ interval, Vött et al. (2011b) suggested that this dating result fits the large AD 551/552 earthquake, mentioned by the contemporaneous Byzantine author Procopius.

The results, which Vött et al. (2011b) believe provide support to their Olympia tsunami hypothesis could be controlled from many points of view. One is that the hypothesis does not fit the Procopius description at all, for the 551/552 large earthquake(s) were reported to occur in the Gulf of Corinth that is far away to the north-east of Olympia. Yet, the tsunami that supposedly associated the 551/552 earthquake(s), according to Procopius occurred in the Maliac Bay, further north-east in Central Greece, having no connection neither to the Corinth Gulf nor to the Ionian Sea. Moreover, the historical earthquake event of 551/552 stands at time distance of 33 years from the lower limit of the geological time window obtained by the radiocarbon dating. The time interval of 33 years equals more

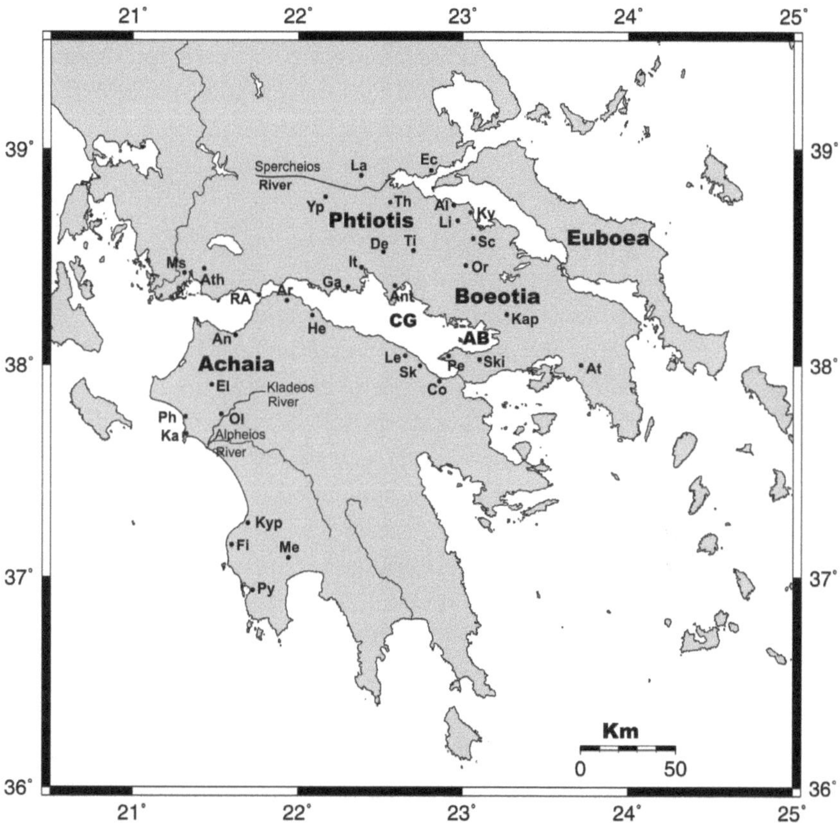

Figure 6.1 ***Geographic areas and localities mentioned in Sections 6.3.2.1–6.3.2.3.*** Key: CG, Corinth Gulf; AB, Alkyonides Bay; RA, Rion–Antirion strait; Al, Alopi; An, Antikyra; And, Ayios (St.) Andreas; Ar, Arachovitika; At, Athens; ATh, Ayios (St.) Thomas; Co, Corinth; De, Delphi; Ec, Echinus; El, Elis; Fi, Filiatra; Ga, Galaxidi; He, Helike; It, Itea; Kap, Kaparelli; Kat, Katakolon; Ky, Kynos; Kyp, Kyparissia; La, Lamia; Le, Lechaion; Li, Livanates; Me, Messene; Ms, Messolonghi; Ol, Olympia; Or, Orchomenos; Pe, Perachora; Ph, Pheia; Py, Pylos; Sc, Scarphe; Sk, Skoutela; Ski, Skinos; Ti, Tithorea; Th, Thermopylae; and Yp, Ypati.

than half the time window itself, which indicates that the date 551/552 falls significantly out of the 2σ-interval of the radiodated time window.

Another point of criticism is that from geological point of view the tsunami is very weakly documented, and this is exactly why Vött et al. (2011b) speak only about a hypothesis. Besides, these authors recognized that there are no historic accounts on catastrophic flooding of Olympia. Finally, it is highly questionable whether even a tsunami of extreme size would be capable to run such a long distance inland, not only to reach ancient Olympia but also to preserve energy to cause destruction there. Such a question could be met by hydrodynamic experiments with numerical tsunami simulations, which is beyond the scope of this section. A rough approximation, however, can be based on the empirical relationship proposed by Hills and Mader (1997) for tsunami inland penetration, d (in kilometer):

$$d = H^{1.33} n^{-2} k \tag{6.4}$$

where H (in meters) is the wave height in the sea shore, n is Manning's n, and $k = 0.06$ is a constant. Manning's n equals 0.015 for smooth terrains such as mud flats and or pastures, 0.07 for densely treed landscapes and 0.03 for areas covered in buildings. For this application I calculated $d = 0.572$ km by considering $n = 0.07$ and $H = 20$ m. Today the distance of Olympia to the closest shoreline is about 18 km, which is extremely larger than the calculated distance, thus weakening further the Olympia tsunami hypothesis. The argument does not change by considering that the shoreline has advanced considerably due to the Alpheios River deposition. The ancient coast at the mouth of Alpheios River was situated only some 3 km inland from its present position.

In a more thorough study of Vött et al. (2011c) it seems that the Olympia tsunami hypothesis has been abandoned but detailed results were presented to support that the coastal zone of Pheia, the harbor of Olympia, was inundated by three distinct tsunami events in the geological past. From geological observations and archeological findings those authors suggested that the most recent tsunami occurred in Byzantine to post-Byzantine period. To correlate the supposed tsunami event with the 551/552 earthquake, Vött et al. (2011c) cited Kraft et al. (2005) who just were based on the hypothesis put forward without documentation by Yalouris (1957, 1960). This is a characteristic example of what in historical seismology we may call "propagation of uncontrolled information without discrimination." To support further their correlation, Vött et al. (2011c, pp. 262–263) stated that *The 551/552 AD earthquake is reported to have been related to a strong seismic sea wave which brought destruction to many coastal cities on the Peloponnese and the Greek mainland* (Guidoboni et al., 1994) and that *written sources are in favour of the year 551/552 AD when a seismic sea wave is reported to have hit the area (p. 280)*. However, one may find no such a statement in Guidoboni et al. (1994) who, *per contra*, quoted carefully only the text of Procopius, which, as we have seen, does not speak about earthquake(s) and tsunami in the Peloponnesian coasts in general but only in Corinth Gulf and Maliac Bay, Central Greece. On the other hand, the large tsunami of AD 365, which is well documented from many points of view, remains as candidate event to interpret the tsunami traces found in Pheia, provided that the tsunamigenic origin of the traces are reliable ones.

The multiple descriptions of paleotsunami events in the Greek side of the Ionian Sea have important consequences for the tsunami hazard and risk assessment. For example, Vött et al. (2007, p. 55) supported that *multiple tsunamigenic influences during the late Holocene show that Aktio headland is exposed to a high tsunami risk. The adjacent NATO airport and nearby centers of mass tourism indicate high vulnerability of the area* ... Since the high occurrence rate of the paleotsunami events that were supposedly documented by geological methods are in direct contrast to the historical documentation of tsunamis in the area, the relevant geological results should be considered with high skepticism when taken into account the evaluation of the tsunami hazard and risk.

6.3.2.3 The Corinth Gulf

In the tectonic rift of the Corinth Gulf, continental Greece, several sediment layers found in two test sites were attributed to prehistorical and historical tsunami events (Kontopoulos and Avramidis, 2003; Kortekaas et al., 2003, 2011). As regards the historical period, the geological results fit well enough to the historically documented destructive earthquake and tsunami events of AD 1402 and 1870 in the central Corinth Gulf and AD 1817 in the western Corinth Gulf (Kortekaas et al., 2011) (Chapter 2). No geological evidence was found so far for the Helike BC 373 and AD 1861 tsunamis, which are historically known to have been caused by strong earthquakes that ruptured the western Corinth Gulf (for geographical references of this section, see Figure 6.1).

In the eastern Corinth Gulf, multiple late-Holocene tsunami landfills were supposedly recorded in the paleotsunami geoarchive at the Lechaion harbor. Namely, Hadler et al. (2013) supported that they were able to recognize that Lechaion was hit by strong tsunami impacts in the eighth–sixth century BC, the first–second century AD, and again in the sixth century AD. Further, on the basis of their findings those authors (Hadler et al., 2013, p. 169) supported that the tsunami hazard for the Lechaion Gulf seems to be strongly underestimated. On the other hand, none of the historical earthquake and tsunami events, which are known in the Corinth Gulf (Papadopoulos et al., 2000) are correlated with the geologically identified tsunamigenic layers in Lechaion. One may support that the layer dated in sixth century AD could be attributed to the 551/552 earthquake(s) described by the Byzantine author Procopius. However, as we have seen earlier, Procopius speaks for only one tsunami event taking place in Maliac Bay, which has no physical connectivity with the Corinth Gulf (Figure 6.1).

As for the other geologically identified tsunami events in Lechaion, one interpretation could be that the historical earthquake and tsunami catalog in Corinth Gulf is incomplete. This is true, but it is noteworthy that in Corinth Gulf both the seismic and tsunami activity decreases gradually from west to east and no strong tsunami occurrence is known in the east sector of the Gulf (Papadopoulos, 2003b). On the other hand, one may doubt about the tsunamigenic origin of the sediment layers. Part of the main evidence in favor of the tsunamigenic origin of those layers is the presence of beachrock-type calcarenitic tsunamites (Hadler et al., 2013). However, several authors, including Mourtzas and Marinos (1994), Pirazolli (2010), and Mourtzas et al. (2013), believe that the beachrock is lithified beach and indicator for sea level changes. This conflict indicates that before considering the level of tsunami hazard based on the paleotsunami record, further geological research is certainly needed with the aim to verify the tsunamigenic origin of the sediment layers found in Lechaion.

6.3.2.4 Western Mediterranean and SW Iberia

The possible contribution of historical tsunami data in the tsunami hazard assessment of the western Mediterranean area, such as in the Spanish and Algerian coasts, has been qualitatively discussed by several authors (Campos, 1991; Yelles Chaouche, 1991; Yelles Chaouche et al., 2009; Roger and Hébert, 2008). In the area of SW Spain, sedimentological and geomorphological imprints of Holocene tsunamis along with historical tsunami record (e.g., the tsunamis of 218–209 BC and AD 1755) were combined and the recurrence period was calculated as ranging from 700 years to 1000 years (Ruiz et al., 2013). In SW Iberia area, high energy, extreme events were identified from sedimentary, paleontological, and geomorphological records and were interpreted as tsunamigenic in origin by Lario et al. (2011) (Chapter 2). Those authors estimated that in SW Iberia extreme wave events have a periodicity of 1200–1500 years. However, this result introduces a very high uncertainty as regards the tsunami hazard assessment since not all of the events considered were tsunamigenic. Matias et al. (2013) used fault models to estimate the recurrence of large, potentially tsunamigenic earthquakes in the Gulf of Cadiz with prospects of application not only in the tsunami hazard assessment but also in tsunami warning procedures.

6.3.3 Statistical and Probabilistic Approaches

Parametric tsunami catalogs offer observational material, which in principle is suitable for tsunami hazard evaluation from statistical and probabilistic approaches. Probabilistic tsunami hazard assessment (PTHA) around the globe were developed mainly after the 2004 Indian Ocean devastating megatsunami (Geist and Parsons, 2006). However, several papers were published in the Mediterranean region before the 2004 megatsunami. Authors that published relevant studies in the EM region and elsewhere as a rule were inspired by the seismological experience in calculating seismic hazard. However, a sensitive issue is the relatively low number of events inserted in the various catalogs as well as the high catalog incompleteness. In seismology, the traditional way to control catalog completeness is to plot event numbers against event magnitudes and to obtain the well-known Gutenberg–Richter (G–R) or magnitude–frequency diagram. Then, incompleteness is adopted for the point where the plot deviates from the G–R line at the left hand-side sector of the diagram. However, incompleteness may also come from the fact that tsunami events escape reporting even in the instrumental period, that is in the last 100 years or so, as the examples of the Calabrian Arc (Graziani et al., 2006), Hellenic Arc (Papadopoulos et al., 2007a, 2010), and Adriatic Sea (Maramai et al., 2007) indicated.

Another problem in tsunami science is the quantification of the tsunami events. In fact, as analyzed in Chapter 1, no adequate tsunami magnitude scale has been introduced so far. In the EM region, only for a very few tsunami events magnitude has been calculated (see Appendix). Therefore, usually tsunami intensity is in use as a proxy of tsunami magnitude. From this point of view the assessment concerns hazard. However, since intensity is an expression of impact, one may argue that the assessment concerns rather tsunami risk than tsunami hazard.

One of the first attempts to apply seismological methodology (intensity–frequency distribution, intensity attenuation) and to calculate tsunami hazard for selected coastal sites of Greece in terms of the most probable maximum expected intensity in a certain time interval was presented by Papazachos et al. (1985, 1986). Tsunami size–frequency relations were also used to calculate tsunami mean repeat times in the Black Sea (Pelinovsky, 1999). Some years later, based on a widely revised tsunami catalog for the Corinth Gulf, Central Greece, the author of this book (Papadopoulos, 2003b) was able to extend the previously mentioned statistical tsunami hazard assessment by including estimates of tsunami Poissonian probabilities (Figure 6.2). Tsunami intensity was used as a proxy to tsunami magnitude. It was found that in time intervals of 100, 175, and 340 years, the probability becomes nearly 100% for the occurrence of at least one tsunami event of intensity equal to or larger than 2, 3, and 4, respectively, in the 6-grade intensity scale

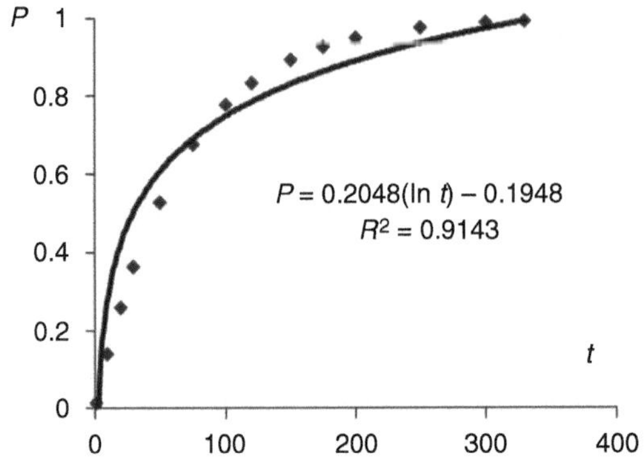

Figure 6.2 ***Poissonian probability function for the occurrence of tsunamis of intensity ≥4 in the Corinth Gulf, Central Greece (Papadopoulos, 2003).***

of Sieberg–Ambraseys. The systematic revision of the tsunami catalog in the eastern Mediterranean basin made it possible to follow the same statistical and probabilistic methodology in the eastern and western segments of the Hellenic Arc (Papadopoulos et al., 2007a, 2010) as well as in the area of Cyprus and the Levantine Sea (Fokaefs and Papadopoulos, 2007).

One of the major obstacles for statistically calculating tsunami hazard parameters, such as the recurrence period of events exceeding a cut-off threshold of size, is the small number of events contained in tsunami catalogs. The relatively low number of events in conjunction with the incompleteness of the tsunami catalog makes it very risky to apply the previously mentioned methodology in other areas of the EM region. To overcome this difficulty, Tinti (1991, 1993) developed an interesting approach and applied it in Italy. The basic idea was to calculate the recurrence of earthquakes from seismicity data, which are more abundant, and then to translate the results to tsunami probabilities based on suitable assumptions. Namely, in each seismogenic region the earthquake potential has been statistically calculated in small cells of equal size, for example, 20′ × 20′. Then, probabilities for tsunami generation were calculated for each cell by taking into account the earthquake size, focal depth, and focal mechanism. The same approach was applied for the tsunami hazard evaluation in the Moroccan coasts (El Alami and Tinti, 1991). The disadvantage of this approach was again the incompleteness of both the earthquake and tsunami catalogs as well as the set of various assumptions that should be made in order to translate seismic potential to tsunami potential. A refinement of the previous procedure was used for probabilistic seismic hazard assessment in Calabria and Sicily, Italy and the estimated tsunami activity, expressed in terms of the number of expected events in a 10,000-year period with runup heights exceeding a given threshold value, was compared with historical tsunami catalog data (Tinti et al., 2005a).

The tsunami hazard in the entire Mediterranean region was evaluated using various qualitative and quantitative independent approaches on the basis of a catalog constructed from the geological and historical tsunami record over the last 3.5 millennia or so (Papadopoulos, 2009). One of them comprised a comparative description of the spatial distribution of tsunami events. It was found that the tsunami activity increases from west to east across the Mediterranean, having its maximum in the area of the Hellenic Arc. With a second approach, the tsunami recurrence was calculated in the entire Mediterranean Sea, Greece, and Italy from empirical intensity–frequency relationships. Another statistical approach was to estimate in a uniform way the level of tsunami potential across the several tsunamigenic zones of the Mediterranean Sea (Figure 2.1) by taking into account both the frequency of occurrence and the weighted event intensity.

A probabilistic estimate of earthquake-generated tsunami hazard for the entire Mediterranean Sea was presented by Sørensen et al. (2012). They estimated the annual probability of exceeding a given tsunami amplitude at any coastal location in the region by applying a Monte Carlo based technique. This PTHA procedure showed that the highest hazard is in the eastern Mediterranean owing to earthquakes along the Hellenic Arc, but most of the Mediterranean coastline is prone to tsunami impact. The probability of a tsunami wave exceeding 1 m somewhere in the Mediterranean in the next 30 years is close to 100%. A Bayesian probability procedure for tsunamis generated by both seismic and landslide sources was presented by Grezio et al. (2012) for the Messina Strait Area (Sicily, Italy).

Recently, Papadopoulos et al. (2014a) reviewed the available historical and geological record and concluded that a list of reliable tsunamis with assigned intensity 6 or larger in the 12-grade Papadopoulos–Imamura tsunami intensity scale may include 44 events across the Mediterranean and its adjacent seas. This list is reproduced in the Appendix. They found that 27 out of 44 tsunami events occurred in the east Mediterranean basin, which turns to a mean tsunami recurrence of 93 years. Respective rates in the western Mediterranean basin, the SW Iberian Margin, the Marmara Sea, and the Black Sea are 227, 2500, 500, and 1250 years. However, such statistics is perhaps underestimated due to possible data incompleteness.

6.3.4 PTHA From Incomplete Tsunami Catalogs

In seismology, catalog incompleteness has been discussed extensively for many years. A methodology, which was developed since the end of 1980s considers earthquake hazard assessment from earthquake catalogs which are incomplete with different incompleteness levels in different time segments of the catalog (Figure 6.3). In this sense, the earthquake catalog may include prehistoric (palaeoseismological), historic, and instrumental data sets. This means that the origin time of some events, for example, prehistoric or historic ones, can be uncertain. In other words, although the origin times of the seismic events occurring in the instrumental period are precise, in the origin times of prehistoric and historic events the date or even the month of occurrence could be missing.

Several relevant papers have been published but the ones that summarize the overall methodology are the ones published by Kijko and Sellevoll (1989, 1992) for the estimation of seismic hazard parameters, like the activity rate, λ, the parameter, β, which is interrelated to the b-value of the G–R (magnitude–frequency) relationship ($b = \beta \log e$), and the maximum magnitude, M_{max}, expected to occur in a target area. In those papers, the mathematical formalism was introduced for the seismic hazard estimation from incomplete data files and the incorporation of magnitude heterogeneity. The methodology was applied successfully in several seismogenic regions, including the Hellenic Arc (Papadopoulos and Kijko, 1991). Several iterations for Hellenic Arc earthquake catalogs with small number of events, n, showed that the results stabilize for $n > 5$. Later on, Kijko (2004) developed further the method for the estimation of the maximum earthquake magnitude (see also in Kijko and Singh, 2011), while more recently Kijko and

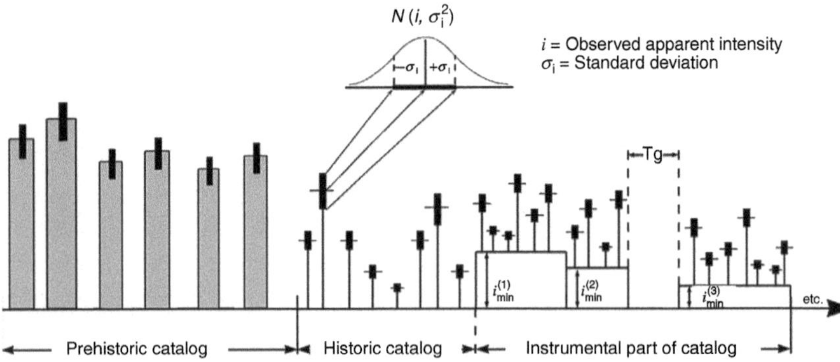

Figure 6.3 ***Schematic illustration of incomplete catalog data that may be used in the seismic but also in the tsunami hazard assessment (Kijko and Sellevoll, 1989, 1992).***

Smit (2012) presented an extension of the Aki–Utsu *b*-value estimator for incomplete earthquake catalogs.

Considering that a tsunami catalog is available, then it is possible to extend the application to the probabilistic tsunami hazard assessment. The methodology is absolutely suitable since tsunami events are quite infrequent compared to the occurrence frequency of seismic events and, therefore, tsunami catalogs are highly incomplete data files with incompleteness strongly varying with time. Again the catalog may include data from palaeotsunami, historic, and instrumental events with uncertain origin times.

However, at the present stage of research no standard method exists for the tsunami magnitude calculation although several methods have been proposed. Therefore, either tsunami wave height, *H*, or tsunami intensity, *K*, are in use to measure the tsunami size. Parameter *H* is a physical feature of the wave, while *K* is an expression of the tsunami impact. As a consequence, wave height *H* expresses the tsunami hazard level. On the other hand, *K* reflects the tsunami risk level. Average tsunami wave height, *H*, observed in a particular coastal segment, is interrelated to the parameter, *S*, defined by Soloviev (1970) as

$$S = \log_2[2H]^{-2} \tag{6.5}$$

Soloviev (1970) considered *S* as a measure of tsunami intensity but it is rather a measure of tsunami magnitude given that *S* is only a physical parameter and does not provide information about the tsunami impact.

PTHA products from incomplete tsunami catalogs can be produced for a coastal site (e.g., Heraklion city) or segment (e.g., north Crete) or zone/region (e.g., the eastern Mediterranean basin) depending on the data available and the purpose of the study. The products of such an analysis include the annual probability of exceedance, the probabilities of exceedance in several time intervals as well as the return period of the wave height or other attribute that may describe tsunami hazard. The PTHA from incomplete tsunami catalogs was tested in Heraklion city, north Crete, which is one of the test sites selected for the EU FP7 tsunami research project ASTARTE. The methodology was applied recently by research teams of the National Observatory of Athens and the Pretoria University of South Africa. Tsunami hazard assessment was considered in terms of average wave height, *H*. The parameters which were calculated included the activity rate, λ, and the *b*-value for the height-frequency relationship. The time flow and the size distribution of the tsunami events were modeled by Bayesian formalism, which takes into account the uncertainty of the tsunami hazard model. The parameters λ and *b* were calculated simultaneously with the

Table 6.1 Tsunami data for the Heraklion test site, Crete

n	*Y*	*M*	*D*	*H*	*K*
1	(−) 1613			10	10
2	1303	08	08	5.0	10
3	1494	07	01	1.0	4
4	1612	11	08	1.0	4
5	1650	10	10	4.0	6
6	1741	02	?	1.0	4
7	1956	07	09	1.5	5
8	2000	04	05	0.5	4

Although assessment of tsunami risk is not considered here, tsunami intensity, *K* (in the Papadopoulos–Imamura 12-point scale), is also listed for reasons of providing information on the impact of the tsunami events considered. The event with time of occurrence −1613 is the LBA tsunami produced by the Santorini volcano caldera collapse. *n*, code number; *Y*, year; (−), BC; *M*, month; *D*, day; *H*, tsunami height in meters.

maximum likelihood approach, while the errors involved in the tsunami size, *S*, were modeled by Gaussian distribution. The range of tsunami size takes the form $<S_{min}, S_{max}>$. The estimation of the maximum tsunami size, expressed in wave height *S*, can be done by following a variety of procedures applied in earthquake catalogs (see Kijko, 2004; Kijko and Singh, 2011, for review and references). The computer program for the tsunami hazard assessment was written by Dr A. Kijko initially in 2012 with the last updated revision completed on June 10, 2013. The tsunami catalog data used are listed in Table 6.1.

In the calculations, the beginning and termination of the tsunami catalog listed in Table 6.1 was conventionally taken on January 1, 1615 BC and December 31, 2000 AD, respectively, that is, the time span of the whole catalog is 3615 years. The number of catalog events is eight. The prior value of *b* was taken 0.50 with standard deviation 0.1. The maximum tsunami wave height, in terms of *S*, was calculated according to the procedure of Kijko–Sellevoll–Bayes (see details in Kijko, 2004; Kijko and Singh, 2011). The model uncertainties for the parameters λ and β were 25%. The introduced threshold for the Soloviev parameter, *S*, is 0.10, while the standard error of *S* was assumed equal to 0.25. The calculations were performed for minimum threshold $S = 0.10$. The maximum observed value of *S* is 2.20 ± 0.10.

The results obtained for the tsunami hazard parameters, for minimum intensity $S = 0.10$, are $\beta = 1.34 \pm 0.02$ or $b = 0.58 \pm 0.09$ and $\lambda = 0.013 \pm 0.005$, $S_{max} = 2.48 \pm 0.24$ (for S_{max} observed 2.20 ± 0.10). The COV(λ, β) is −0.270. Figures 6.4–6.6 illustrate the annual probability of

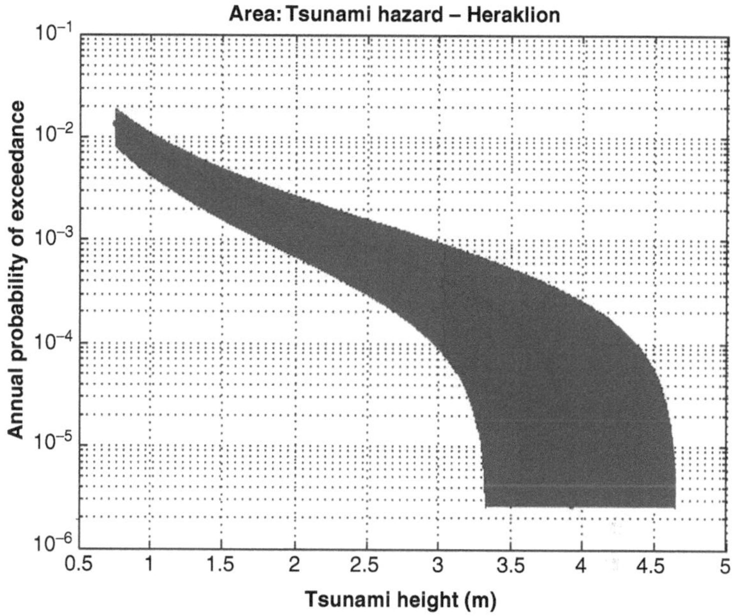

Figure 6.4 ***Annual probability of exceedance of the Soloviev tsunami wave height, S, as determined by formula (6.5).*** *(by A. Kijko, G.A. Papadopoulos & A. Smit, unpublished report, see text for details.)*

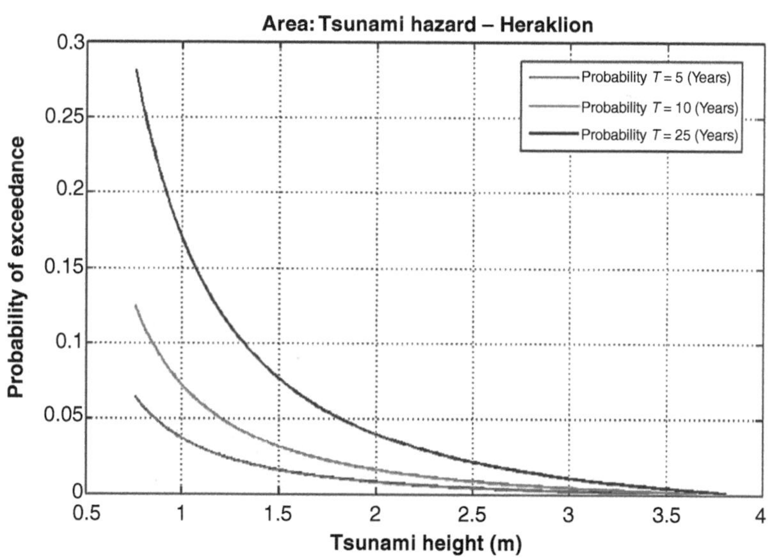

Figure 6.5 ***Probabilities of exceedance, in time intervals of 5, 10, and 25 years, of the Soloviev tsunami wave height, S, as determined by formula (6.5).*** *(by A. Kijko, G.A. Papadopoulos & A. Smit, unpublished report, see text for details.)*

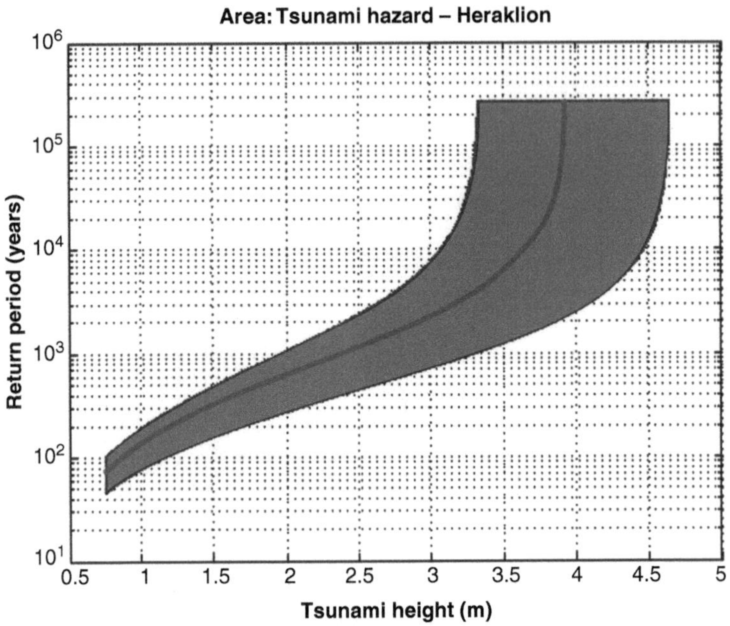

Figure 6.6 ***Return period (in years) of the several levels of the Soloviev tsunami wave height, S, as determined by formula (6.5).*** *(by A. Kijko, G.A. Papadopoulos & A. Smit, unpublished report, see text for details.)*

exceedance, the probabilities of exceedance in time intervals of 5, 10, and 25 years as well as the return period (in years) of the several levels of the Soloviev parameter *S*.

The probabilistic hazard assessment from incomplete catalog data, which was successfully tested with incomplete earthquake catalogs since the 1990s, was effectively applied in the Heraklion test site (Crete Island) from incomplete tsunami catalog containing small number of tsunami events. One may understand that the same methodology could be applied for tsunami risk assessment provided that data of tsunami intensity or other impact quantity are available.

6.3.5 Scenario-Based Hazard Assessment

A methodology for tsunami hazard assessment that gains ground in the last years is based on the numerical simulation of tsunami waves that are synthetically produced by tsunami scenarios. Two main types of scenarios could be considered. The first refers to the so-called "Realistic Scenario," which is based on the reproduction of a causative source, for example, an earthquake or a landslide, with parameters as close as to the known, or inferred, parameters of the source which is known that has generated a tsunami in the past. An example of a realistic scenario application, based on the AD January 11, 1693 earthquake-generated catastrophic tsunami in Sicily can be found in the paper by Tinti and Tonini (2013). The other type comprises the "Worst-Case Credible Scenario" or "Maximum Credible Scenario (MCS)." An MCS may consider a tsunami source that has the maximum possible size, for example, maximum earthquake magnitude or fault. In this sense the MCS is based either on a real causative event known from the past, for example, earthquake, or on a virtual event that could be generated from geotectonic or geological structures that have the potential to produce tsunami(s) in the future. In any case, to secure credibility of a scenario it should take into account historical, geotectonic, and geological information as well as a careful selection of potential tsunami sources on the basis of such types of information.

More than one realistic and/or MCS's could be considered for the tsunami hazard assessment in a particular coastal area. In addition, the technique of aggregating results of more than one scenarios is possible to apply while more than one different types of sources (seismic, landslide, volcanic) may be inserted in the scenarios and the aggregations. Aggregation may consider the field of water elevation or other hydrodynamic variables, such as flow depth, water velocity, minimum elevation, and others. Such techniques are capable to provide inundation maps in the coastal zone of interest. These maps are quite useful for the tsunami vulnerability and risk assessment as well as for undertaking risk mitigation actions, for example, designing evacuation zones. If combined with probabilistic assessments the scenario-based techniques may also yield probabilistic or statistical return times of tsunami inundation of certain levels.

An example of aggregated data comes from the study (Armigliato et al., 2013) that was performed by the tsunami team of the University of Bologna, Italy, for the island of Rhodes, Greece, which was the test site of the tsunami project NEARTOWARN supported (2012–2013) by the DG-ECHO, EU. In that study, several scenarios were selected and numerically modeled. The scenarios included local earthquake (LEQ), regional earthquake (REQ), and local landlside (LS). Figure 6.7 shows the aggregated maximum water elevation field, while Figure 6.8 shows for every node which scenario has determined the maximum elevation. The LS and LEQ scenarios are dominant in the NE of the island, while the REQ scenario prevails in the rest. The conclusion is that the highest tsunami waves can be observed in Rhodes as a consequence of an earthquake in the Hellenic Arc, while local sources contribute less, though are important for the town of Rhodes and the neighboring area.

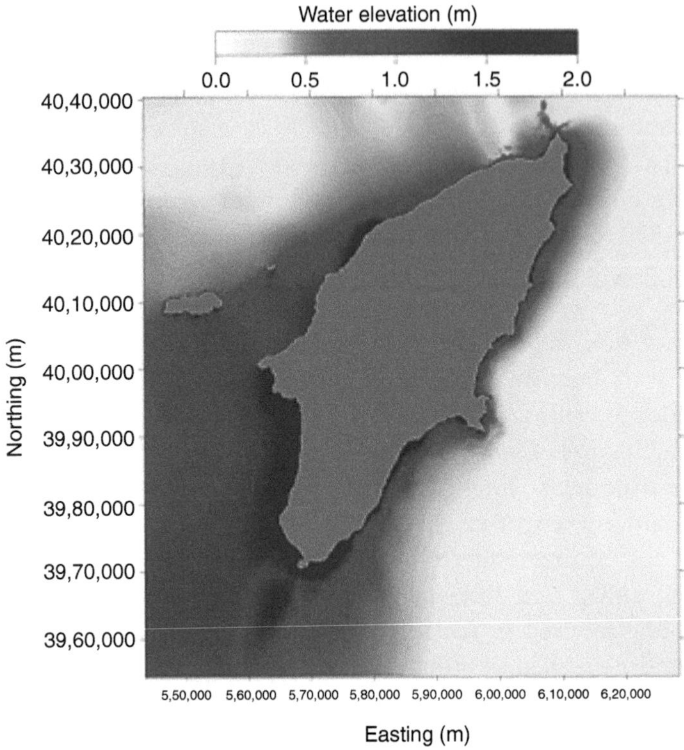

Figure 6.7 ***Aggregated maximum water elevation field for Rhodes Island (Armigliato et al., 2013).***

In the Mediterranean region several efforts were made in the last years for the assessment of tsunami hazard based on scenarios. One of the first efforts were published by Tinti et al. (2005) who developed simple scenarios of earthquake-generated megatsunamis in four seismogenic areas in the western, central, and eastern sectors of the Mediterranean basin. From the numerical simulation of such megatsunamis it was found that both local and remote coastal zones are threatened, the local being attacked in about within 15 min from the tsunami generation time. In 1 h or so, tsunamis cross the basin and hit the remote coasts opposite to the source, namely, France and Liguria (Italy) for the Algerian source, Greece, Tunisia, and Libya for the eastern Sicily source, Italy and northern Africa for the Hellenic sources. However, historically only the last case in known that has happened.

The possible impact on the UK coastline of the mega 1755 Lisbon tsunami was investigated by Horsburgh et al. (2008). Six different initial conditions were used but only one of the source models, an assumed earthquake

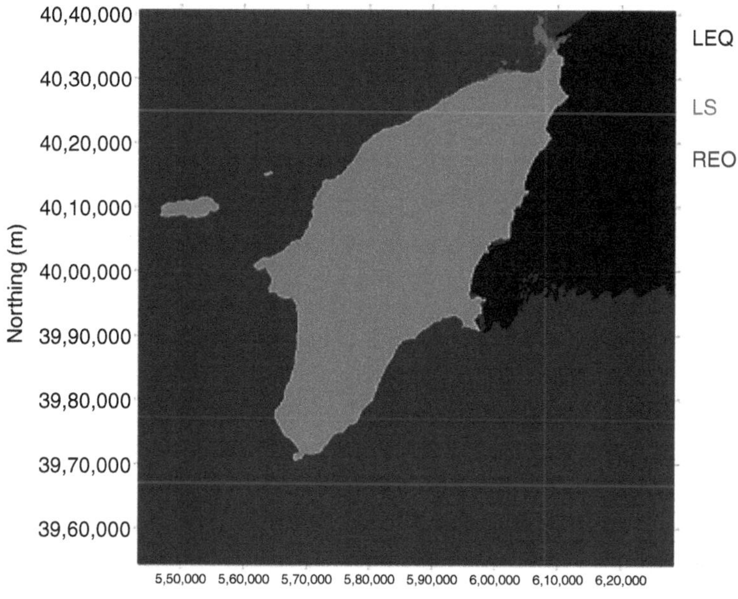

Figure 6.8 ***Influence of the different scenarios in determining the field plotted in Figure 6.7 (Armigliato et al., 2013).***

magnitude of M_w8.7 with a zonal fault orientation, resulted in significant wave heights at the UK coastline. In spite that only a fraction of tsunami energy reaches the northwest European shelf, significant local reamplification due to wave interactions and resonance on the continental shelf was found. The maximum elevations obtained were comparable to severe winter storm conditions, but with extreme local variability in the tsunami amplitude. The results suggested that the impact of any repeat of this event would be very sensitive to the precise location and orientation of the source deformation, as well as by complex topographic interactions on the shelf. Horsburgh et al. (2008) noted that the uncertainties arising from the combination of source orientation and bathymetric interaction suggest that any assessment of risk, for places where tsunamis are likely, should consider a large ensemble of initial conditions.

In the Adriatic Sea domain, Paulatto et al. (2007) calculated several synthetic mareograms applying to the shallow water basin case both the theory of modal summation by Panza et al. (2000) and the theory of Green's function by Yanovskaya et al. (2003). The first was applied to the case of tsunamis generated by an offshore source, the second to the case of tsunamis generated by an inland source. Synthetic mareograms were calculated by varying parameters such as earthquake magnitude, focal depth, and water layer thickness, in order to estimate the expected values of tsunami maximum amplitude and arrival time, in the whole Adriatic basin, for the selected scenarios.

Another approach was followed by Tiberti et al. (2008) for the tsunami hazard assessment in the Italian coast of the Adriatic Sea. They calculated the expected impact of a large set of tsunamis resulting from potential earthquakes generated by major fault zones. The approach merges updated knowledge on the regional tectonics and scenario-like calculations of expected tsunami impact. For each one out of six selected potential source zones, they determined a Maximum Credible Earthquake and the associated Typical Fault, described by its size, geometry, and kinematics. Then, the Typical Fault was allowed to float along strike of its parent source zone. All tsunamis that could generate were simulated based on the solution of the nonlinear shallow water equations through a finite-difference technique. For each run the wave fields at specified simulation times and the maximum water height field (above mean sea level) were calculated, while travel-time maps, and maximum water height profiles along the target coastline were generated. Maxima classified in a three-level code of expected tsunami threat. It was found that the southern portion of Apulia facing Albania and the Gargano promontory are especially prone to the tsunami threat. They also found that some bathymetric features are crucial in determining the focalization–defocalization of tsunami energy.

The same Maximum Credible Earthquake approach was applied by Lorito et al. (2008) to assess potential tsunami threats for Southern Italy from three potential source zones located at short (Southern Tyrrhenian thrust belt), intermediate (Tell-Atlas thrust belt), and large distances (Western Hellenic Arc). It was found that there is a highly variable impact for tsunamis generated by the different source zones. For example, a large Hellenic Arc earthquake will produce a much higher tsunami wave (up to 5 m) than those of the other two source zones (up to 1 m). Therefore, Lorito et al. (2008) concluded that tsunami scenarios for Mediterranean Sea countries must necessarily be computed at the scale of the entire basin.

For the tsunami hazard assessment on the southern coast of Spain, Alboran Sea, an indirect statistical method combined with scenario-based numerical models was proposed (Gonzalez et al., 2010). The method included tectonic and seismicity analysis, tsunami source model, generation of a numerical data base of tsunami events using a numerical model, probabilistic model based on Monte Carlo simulations in order to generate a synthetic tsunami catalog and a multidimensional interpolation method applied to combine the numerical data base with the synthetic catalog aiming to produce inundation maps. The purpose of that study was to establish tsunami wave elevation at the shoreline versus the return period curves for different locations along the Alboran Sea coast. It was concluded that the tsunamis generated in the Alboran basin have a medium to low hazard, with the most important elevations in the Malaga, Adra, and Melilla areas.

For the city of Catania, Sicily, the estimation of the tsunami hazard was performed by using the technique of the Worst-case Credible Tsunami Scenario Analysis (Tonini et al., 2011). After building scenarios proposed on the basis of tectonic considerations and of the largest historical events that hit the city in the past, those authors combined all the information deriving from single scenarios into a unique aggregated scenario that can be viewed as the worst virtual scenario. Scenarios were calculated by means of numerical simulations on computational grids of different resolutions, passing from 3 km on a regional scale to 40 m in the target area. La Plaia beach resulted to be the area most exposed to tsunami inundation, with inland penetration up to hundreds of meters. The harbor turns out to be more exposed to tsunami waves with low frequencies: in particular, it was found that the major contribution to the hazard in the harbor is due to a tsunami from a remote source, which propagates with much longer periods than tsunamis from local sources.

Studies relevant to the concept of scenario-based tsunami hazard assessment were also published for the Corinth Gulf for seismic tsunami sources by Stefatos et al. (2006) and Tselentis et al. (2006), and for landslide sources by Tinti et al. (2007), for the Ionian Sea (Tselentis et al., 2010), and the southeast Aegean Sea (Mitsoudis et al., 2012) for seismic sources as well as for SW Peloponnese, Greece, for local and distant seismic sources and the distant volcanic source of Santorini (Papadopoulos et al., 2014b).

One of the problems associated with the scenario-based tsunami hazard assessment is the uncertainty, both *epistemic* and *aleatory*, which is involved in the several methods proposed. Epistemic uncertainty refers to limited knowledge we may have about the system, modeled or real, that we are interested in. However, the aleatory uncertainty comes from a random process given that it refers to the inherent uncertainty due to the probabilistic variability. Therefore, this type of uncertainty is irreducible, in that there will always be variability in the underlying variables. Only little research has been done so far as regards the reduction of the epistemic uncertainty involved in the various approaches, which are in use for the statistical and probabilistic tsunami hazard assessment. Nevertheless, one may consider that it is treated by the incorporation in the various tsunami models of more and more information, for example, seismotectonic, geological, geophysical, bathymetric, etc.

In a very recent paper, Lorito et al. (2015) considered that for earthquake-generated tsunamis a very large number of tsunami simulations are typically needed to incorporate in Seismic PTHA the full expected variability of seismic sources (the aleatory uncertainty). To reduce their number they proposed a probabilistic filtering procedure for selecting a reduced set of seismic sources and, eventually, to calculate "nonlinear" probabilistic inundation maps at the target site, using only the selected sources. They found that the selection of the important sources needed for approximating probabilistic inundation maps can be obtained based on the offshore maximum tsunami amplitude values only. The filtering procedure is semiautomatic and can be easily repeated for any target sites. Taking as a case study the Hellenic Arc, it was found (Lorito et al., 2015) that this approach allows for a 75–80% reduction of the number of the numerical simulations needed, while preserving the accuracy of probabilistic inundation maps to a reasonable degree.

6.4 TSUNAMI VULNERABILITY ASSESSMENT IN THE EM REGION: QUALITATIVE AND QUANTITATIVE APPROACHES

Post-event collections of tsunami impact data are of critical importance for the development and validation of vulnerability and risk assessment methodologies. However, earlier methodologies initially focused mainly on damage assessment, the concept of vulnerability being rather neglected, while vulnerability was approached in more recent studies, which included the physical and socioeconomic dimensions of tsunami impacts. In particular, after mainly the Indian Ocean tsunami on December 26, 2004 and the March 11, 2011 Tohoku-oki tsunami in Japan several studies focusing particularly on the vulnerability of buildings to tsunamis were published. Currently, there are two main classes of vulnerability analysis applied to tsunamis and other geohazards that can be grossly denoted as qualitative and quantitative approaches.

6.4.1 Early Studies and the PTVA Model

One of the earliest studies was that of Reese et al. (1968) who presented the results of the damage survey performed after the 1960 tsunami at Hilo,

Hawaii. They included observations on the height of wave, the structural damage caused and the pressures acted on structures from the hydrodynamic force of the wave. The effect of floated and washed away timbers as well as oil spread due to tsunamis was investigated by Goto (1990) who developed and applied numerical models for the Miyako Bay in Japan. The maximum impulsive force on structures was examined by Matsutomi (1991). For that purpose, variations of the apparent inertia coefficient and the impact duration time of timbers due to the opening ratio and the size of timbers were examined experimentally. The study concluded that the impulsive force of timbers is not negligible compared with that of bores themselves. Shuto (1991) provided a summary of some old approaches for tsunami vulnerability assessment pointing out the strong connection of structural effects on the resistance against tsunamis. Namely, the author examined the relation between losses of lives and residences based on observations of big past tsunami events in Japan and the damage percentage of houses over tsunami height and the drag force, and the damage percentage of fishing boats over tsunami height.

In his study for Grays Harbor, Washington, and Lima, Peru, Preuss (1991) defined vulnerability by considering tsunami as a system rather than a single physical process. In that study the author introduced the assumption that the physical threat, including inundation, strong currents, and the potential for ground subsidence, should be correlated with land use characteristics. The methodology used consisted of four components: (a) define characteristics and dimensions of direct tsunami threat (direction of energy, wave height runup and arrival time); (b) define vulnerability patterns, that is, land use and population distribution patterns; (c) identify secondary hazards, that is, subsidence/ground failure, battering, fire, potential toxic release; and (d) determine microzonation of risk reduction, which includes land use zonation and preparedness and evacuation district.

In the European–Mediterranean region (EM), one of the first relevant studies performed within the frame of the GITEC-TWO EC Tsunami Project (1996–1998), considered tsunami risk as expressed in formula (6.2), where vulnerability is one of the main three parameters involved (Papadopoulos and Dermentzopoulos, 1998). With an application in a 6-km long coastal segment of the test site of Heraklion, the capital city of Crete Island, Greece, vulnerability was considered and mapped in a series of thematic maps in a semiquantitative way for a variety of assets, such as soil foundation conditions, land use/land cover types, road network, functions, and lifelines, socioeconomic, and population parameters. The worst-case scenario, based on past tsunami history, predicted a 6-m high tsunami attacking the coastal zone of Heraklion. The approach developed by those authors included also assessments of tsunami impact and recommendations for reducing tsunami vulnerability and finally risk. Looking back from a considerable time distance, the study of Papadopoulos and Dermentzopoulos (1998) was a premature realization of the vulnerability analysis recommended later by the EC-Working Paper (2010) which includes: identification of elements and people potentially at risk (exposure), identification of vulnerability factors/impacts (physical, economic, environmental, social/political), assessment of likely impacts, analysis of self-protection capabilities reducing exposure or vulnerability).

Some years after the Heraklion pilot study by Papadopoulos, the approach was further developed as described in the publications of Papathoma and Dominey-Howes (2003) and Papathoma et al. (2003) based on the PhD Thesis of Papathoma (2003). The main developments included the introduction of the concepts of (a) the ***relative vulnerability index (RVI)***, based on several structural criteria for buildings and (b) the human vulnerability (HV) expressed as

$$\mathrm{HV} = \mathrm{BV} \times P\,(BV \text{ is building vulnerability and } P \text{ is Population}) \quad (6.6)$$

The application of the so-called ***PTVA-1 model*** (Papathoma Tsunami Vulnerability Assessment, version #1) was performed in Heraklion, Crete, as well as in a part of western Corinth Gulf, Central Greece, with the inundation zone being considered between the coastline and a certain ground relief contour based on probability studies of historical tsunamis. For example, the contour of 5 m was selected for the Heraklion test site. Thematic maps of BV and HV were displayed in a GIS form.

6.4.2 New Versions of the PTVA Model – Other Approaches

The big tsunami of December 26, 2004, that directly affected 16 nations around the Indian Ocean and indirectly the entire globe, influenced drastically the development of tsunami science and also engineering. This is also the case of tsunami vulnerability assessment thanks to the post-tsunami damage data that are increasingly available from 2004 onward. Abundant data became available after other recent, destructive tsunamis, such as the ones of July 2006 Java and of February 2010 Chile tsunamis, but mainly after the devastating Tohoku-oki, Japan, tsunami on March 11, 2011. As a consequence, over the past decade work has increasingly focused on the development of tsunami-building vulnerability assessment models (Tarbotton et al., 2015). It has been realized that the capacity of a building to withstand the forces imposed by a tsunami depends on a variety of attributes, including its structural design, construction material, foundation type and ground floor characteristics (UNESCO-IOC, 2011). For this reason, most recent approaches are based on empirical functions due to the complex nature of buildings. However, the further development of the PTVA-1 model was not neglected.

In fact, a first line of development was the revision and validation of the PTVA-1 model. Douglas (2007) showed how the lack of post-event observations affected the early models. According to Dall' Osso et al. (2009) the PTVA-1 model was developed before 2004 in the absence of robust, well-constructed, and validated building fragility models for assessing the vulnerability of buildings to tsunami. In view of this reality, Dominey-Howes and Papathoma (2007) upgraded the model to version #2 (PTVA-2) and validated using data collected from field surveys at the Maldives affected by the 2004 big Indian Ocean tsunami. The PTVA-2 model was also tested by Dominey-Howes et al. (2010) in the Cascadia subduction zone (Seaside, Oregon, USA) in relation to the large tsunami of AD 1700. In that study the PTVA-2 model was coupled with a probabilistic tsunami hazard assessment and probable maximum losses were calculated for a 1:500 year tsunami inundation.

Dall' Osso et al. (2009) presented an enhanced version (PTVA-3) of the model that took account of a new understanding of the factors that influence BV and significantly introduced the use of the Analytic Hierarchy Process (AHP) for weighting the various attributes in order to limit concerns about subjective ranking of attributes in the original model. The application was made for Maroubra, Sydney, Australia. In the frame of the EC-FP6 TRANSFER Tsunami Research Project, the same model (PTVA-3) was applied and validated for the assessment of the vulnerability of buildings to tsunamis in the volcanic archipelago of the Aeolian Islands, Italy, and specifically for the islands of Stromboli and Panarea affected by the December 30, 2002 local but powerful tsunami wave (Dall' Osso et al., 2010).

The advantage of the initial PTVA model and its later versions is that it provides estimations of the vulnerability levels for coastal zones in the lack of building damage data from real tsunami attacks. However, the inundation area is only roughly considered from historical data, while tsunami hydrodynamic parameters, such as flow depth or current velocity, are not taken into account.

These problems can be solved with the performance of inundation numerical modeling for tsunami scenarios (Tinti and Armigliato, 2003). This approach was implemented in the tsunami vulnerability study performed by Omira et al. (2010) for the Casablanca harbor and surrounding area in the frame of the EC-FP6 NEAREST Tsunami Research Project. Those authors used a combination of field survey data, GIS, and tsunami inundation numerical modeling, thus linking tsunami hazard, expressed by wave flow depths, with tsunami vulnerability. In this case, the large Lisbon tsunami of November 1, 1755 was modeled as the worst-case tsunami scenario. In addition, this study deviated from PTVA model in that it distinguished between the "main criteria" and the "intrinsic attributes" that influence the building vulnerability to tsunamis. A similar approach was followed for the tsunami impact and vulnerability assessment in the harbor area of Tangier, Morocco (Benchekroun et al., 2013).

Other numerical approaches were also tested to investigate the resistance to tsunami attacks of buildings or critical facilities. For example, Koutitas and Karambas (2005) presented a computational model consisting of the nonlinear shallow water equations in the 2DH space, in the presence of an embedded typical building. The normalized integral of the hydrodynamic loads distributed on the building surface in the direction of the wave propagation is plotted against the normalized length of the structure for various building walls configurations. The importance of the building orientation with respect to the wave propagation and the importance of the openings in the walls were revealed by means of a number of numerical experiments. Cruz et al. (2011), employed the JRC tsunami propagation and inundation code HyFlux2 to describe the potential consequences of two historical tsunami scenarios and their impacts on an oil refinery in Sicily.

6.4.3 Tsunami Vulnerability Assessment Based on Fragility Functions and Damage Curves

6.4.3.1 Fragility Curves and Damage Curves

The utilization of post-tsunami damage data being available over the past decade have contributed greatly to the development of ***empirical vulnerability functions***. In published studies there are two main types of vulnerability functions: the fragility functions and the damage curves. A ***fragility function***, also known as a ***fragility curve***, is a probabilistic vulnerability model that describes the conditional probability that a damage state will be reached or exceeded for a given hazard level (Singhal and Kiremidjian, 1996; Choi et al., 2004; Peiris and Pomonis, 2006). ***Damage curves*** typically represent a structure's damage response as an index or a percentage, referring to the overall level of damage that the building would incur (e.g., 80% damaged). Unlike damage curves, which typically describe the average damage response of a building, fragility functions describe a range of possible damage outcomes and their associated probability of occurrence. This provides the means of describing the damage response of complex structures for which there are large uncertainties in the vulnerability of the components and/or in the loads that are imposed on them (see review in Tarbotton et al., 2015). Reese et al. (2007), Dias et al. (2009), Koshimura et al. (2009a), and Leone et al. (2011) applied such approaches collecting data during post-tsunami field surveys and established the link between observed damage level and tsunami flow depth.

In the aftermath of the March 11, 2011 Great East Japan tsunami, Suppasri et al. (2013) constructed fragility functions using least-squares regressions derived with the use of data for more than 250,000 structures damaged or destroyed. Extended details upon damage level, structural material, number of stories per building and locations were utilized. They also calculated the cumulative probability of damage occurrence using the statistical method described in Koshimura et al. (2009b). Suppasri et al. (2013) showed probability damage results for different buildings classified by their construction materials and number of stores as a function of the flow depth. The results demonstrated very clearly the effect that the mentioned factors have on building damage level.

In the frame of the EC-FP6 SCHEMA Tsunami Research Project, the DamASCHE GIS tool was developed (Valencia et al., 2011) and applied in several test sites of the European-Mediterranean coastal zones by several partners of the project. The DamASCHE GIS tool was initially based on data from Banda Aceh (Indonesia) after the 2004 Indian Ocean tsunami. The tool overlays damage curves onto an inundation map to estimate the damage level. A combined field-remote survey methodology was used, which was capable of classifying building typology and damage state into multiple categories both on the ground (e.g., field surveys) and by inspecting satellite imagery (remote surveys). Building damage was classified into six discrete states ranging from D0 to D5, while building typology was classified into four categories (A, B, C, and D). The tsunami demand parameter was determined by fitting a representative flood depth surface onto available field data using 2D interpolation techniques. Damage curves (Figure 6.9) were developed for the four building typologies by aggregating the data into flood depth bins and considering the mean damage response within

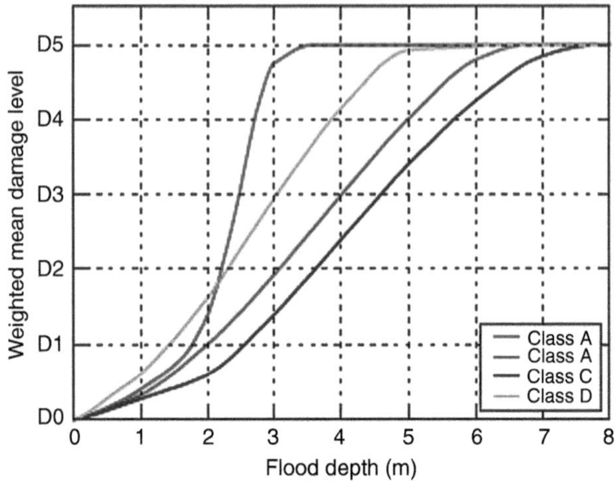

Figure 6.9 ***The average damage curves developed by Valencia et al. (2011).***

each bin interval. The damage curves were taken to be the "enveloping curve" that represented the upper confidence range of the mean damage values calculated at each bin.

The same methodology was applied in several test sites of the SCHEMA Project: by Ranguelov et al. (2010) in Balchik, Bulgarian Black Sea; Atillah et al. (2011), in the Rabat-Salé, Morocco; as well as for Mandelieu, French Riviera in western Mediterranean Sea; Setúbal, Portugal and Catania, Italy (Deliverables 4.3, 4.4, and 4.5, respectively in http://www.schema.project.org/). Omira et al. (2013) investigated the effectiveness of sea-defense structures in preventing/reducing the tsunami overtopping as well as evaluating the resulting tsunami impact at El Jadida, Morocco.

For Padang, an Indonesia case study, Taubenböck et al. (2008), used high-resolution satellite data to assess tsunami vulnerability in urban areas based on building sizes, built-up density, number of buildings, and their location and population. It was proven that remote sensing data and methods in combination with GIS enable a highly detailed classification of vulnerability patterns and their quantification. The potential and limitations of remote sensing techniques (geodata) in vulnerability assessment were investigated by Römer et al. (2012). Their research showed that remote sensing can be used to spatially extrapolate field data on socioeconomic or ecological vulnerability collected in the field, to regionalize exposure elements and assets and to predict vulnerable areas. A similar approach was presented by Mas et al. (2012) for the 2012 Chilean tsunami. In their study, Mas et al. (2012) introduced a practical method suitable when there are limitations on available data for numerical simulation or damage evaluation from surveys: they classified the damage to houses using surveyed data of inundation depth and visual inspection of satellite images of Dichato, Chile and developed tsunami fragility curves.

Gokon et al. (2014), focusing to the areas affected by the 2009 Samoa earthquake and tsunami reproduced the tsunami inundation by numerical modeling and validated the results with field surveys and waveforms from DART gauges. They compared satellite images in order to investigate building damages. Finally they integrated the tsunami features and building damage using GIS and developed fragility functions on statistical analysis as a tool to evaluate structural vulnerability. Those authors pointed out the need to improve their method in terms of taking into consideration floating debris and scourings, building types, as well as age of buildings or floors.

6.4.3.2 Uncertainties Involved

Approaches based on fragility curves and damage curves appear promising for the realistic tsunami vulnerability assessment. However, there are still several problems that make difficult the generalized application of such approaches. One issue is that in most cases such curves are based on empirical data from one location and one event. Also, the structures examined are typically dissimilar to those found in other places. In addition, undamaged structures are often not included which biases the data to damaged structures, which results in overestimating losses. Such critical problems were considered by Dale and Flay (2006) who described an interesting approach directed at assessing the risk posed by tsunami hazard to Australian communities. They proposed to replace curves based on observed damage from a number of past tsunami events by future curves based on an engineering approach. This requires a generalized hazard definition, an engineering model of the structure of interest, and a costing module to convert damage scenarios to restoration coasts. According to Dale and Flay (2006), the generalized hazard definition is a way of generalizing the complex behavior of fluid flow around (and through) a structure and defining the resultant loads on the structure.

Although some design guides exist that contain methods for designing for wave forces, including tsunamis (e.g., the FEMA 55 guide, see FEMA, 2000; the USACE Technical Note III-29, see USACE, 1990), there are a number of reasons it may be problematic to use these guidelines to estimate tsunami wave loads on a structure (Dale and Flay, 2006). Therefore, hazard transfer parameters should be chosen to link between hazard modeling and vulnerability. Water depth and water velocity may be used to describe the hazard at a particular location as an output of the inundation model. Once loads on the structure have been defined an engineering model of a structure can be used to assess damage outcomes. Finally, an automated costing will take the damage outcomes from the engineering model of the structure and cost the repair.

It is of interest that the usefulness of the methodology proposed by Dale and Flay (2006) could be seen from the point of view of the findings reached by Gardi et al. (2011) who examined the uncertainties associated with the process of tsunami damage assessment on buildings. This examination, which was performed within the frame of the EC-FP6 SCHEMA Tsunami Project, concluded that a variety of sources of errors/uncertainties can be introduced at several steps of the damage assessment process. Errors can be introduced when collecting post-disaster observations owing to different types of instruments/methods, water marks considered, tide correction, etc. Much uncertainty comes from difficulties in identifying and characterizing the potential tsunami sources as well as from the numerical modeling and Digital Terrain Models employed and the cartographic processes. Another source of uncertainty is that usually damage functions link the mean damage level on buildings with the maximum water elevation measured in the field without considering other tsunami parameters such as flow velocity.

A comparative review and summary of best practice regarding the use of empirical vulnerability functions for buildings against tsunami impact can be found in the recent paper by Tarbotton et al. (2015). It was found that at present, these functions are highly varied with respect to the building typologies and damage states that they represent, making it difficult to apply them effectively in tsunami vulnerability and loss assessments. Categorization and direct comparisons of existing functions showed that the variation in previous work is due to significant differences in the accuracy of the hazard assessment techniques that are used, inconsistent and poorly defined damage state and building typology classification systems, the use of error-prone and incorrectly applied statistical methods and the use of post-event data that is not representative of the building damage occurring in study area. The recommendations of Tarbotton et al. (2015) are discussed in Section 6.4.6.

6.4.4 The Role of Environmental Factors

The examination of environmental factors, such as coastal vegetation and topography, has shown the important role that such factors play in assessing tsunami vulnerability. For example, coastal forest may decrease or increase tsunami damage. Shuto (1987) pointed out that coastal forest may increase tsunami damage since when huge tsunamis attack the forest, the trees would be destroyed and washed away. Then, a serious secondary damage effect to buildings could be produced by the floating trees. Harada and Imamura (2005) examined the hydrodynamic effect of coastal forest on tsunami hazard mitigation. From numerical simulations it was found that an increase of forest width can reduce not only the inundation depth but also the current and hydraulic force behind the coastal forest. On the other hand, Imai and Matsutomi (2005) examined the 1998 Papua New Guinea tsunami and evaluated the reducing efficiency of tsunami energy caused by the coastal vegetation. They showed that the inertia force reaches 50% of the maximum drag force, FD, at the early stage of inundation flow and that FD, as well as the linear wave making resistance force are dominant at the quasisteady state after the early stage. Also, it was found that both the drag coefficient and the mass coefficient decrease as vegetation density increases.

The effects of complex topography on tsunami runup and inundation distance was examined by Sim et al. (2013). They showed that increasing dunes spacing could not significantly affect inundation distance. However, if the height of sand dunes is of the same order of magnitude as the incoming tsunami wave and the gaps between the dunes are large enough, successful tsunami mitigation could also be possible. Wong (2009) in his study discussed the implications to the integrated coastal management of issues like the loss of land, the safety zone in conflict with the buffer zone, the need for better integration of livelihoods restoration with the rehabilitation of coastal ecosystems and the inclusion of tsunami mitigation in the larger framework of climate change mitigation.

6.4.5 Time-Dependent Vulnerability

From the review presented previously it comes out that several types of tsunami vulnerability could be considered, depending on the elements exposed to the tsunami hazard. The list is long and among others may include:

- Vulnerability of communities (villages, towns, cities) in general
- Structural and other vulnerabilities of buildings, infrastructures, critical facilities
- Vulnerability of population/target groups/individuals
- Indoor/outdoor vulnerability of people and property
- Vulnerability of vessels of variable size as well as of coastal and offshore facilities (e.g., fisheries etc.)

However, the characteristics of the several elements exposed to tsunamis change with time and, therefore, vulnerability changes too. In a short-term sense, vulnerability varies, for example, in a 24-h mode, due to the social and economic cycle of daily life. On a weekly scale the differences in everyday life between workdays and weekends are also well known. On a seasonal scale, strong changes in vulnerability are expected, for example, due to seasonal variation of tourism. Such time-dependent components of the tsunami vulnerability have been noted (Papathoma et al., 2003) but were not considered in the several models proposed for the tsunami vulnerability assessment. As a consequence, practically speaking, vulnerability is not a static, time-independent variable but a dynamic, time-dependent one.

Further, in the long-term sense, the characteristics of the several elements exposed to the tsunami hazard do not remain unchanged. On the contrary, all factors that control the tsunami vulnerability change, for example, built environment, population, land use/land cover, and other environmental factors and so on. For example, as regards the built environment, in earthquake-prone regions, the characteristics of the reinforced concrete buildings change in time with the revision of the antiseismic building codes. This has been taken into account in the seismic vulnerability assessment of building (Kontoes et al., 2012) under the assumption that reinforced concrete buildings constructed by following recent versions of the codes are more earthquake resistant as compared to the ones constructed by following earlier versions of the codes. For example, in Greece the earlier version of the national building code was put forward in 1959, while the last one was established in 2000. A similar situation is realized in Italy and elsewhere.

In tsunami vulnerability assessment, reinforced concrete buildings are generally considered as being of low vulnerability as compared to unreinforced buildings (Papathoma et al., 2003 and later versions of the PTVA model). Although this practice is correct, it is still far behind of the practices applied in earthquake engineering mainly due to the fact that building codes against tsunami loads are still under investigation.

The concluding remark here is that the models developed so far for the tsunami vulnerability assessment are time-independent or static models. A challenging prospect is to go ahead with time-dependent or dynamic models that would incorporate factors that make vulnerability to change on a daily, weekly, seasonal, or even longer scale of time. Then, respective time-dependent models for tsunami risk assessment will emerge.

6.4.6 Summary and Evaluation of the Models for Tsunami Vulnerability Assessment

Reviewing this section, and the several approaches used to assess tsunami vulnerability, one may conclude in a number of critical points that should be taken into account as regards applications not only in the ongoing ASTARTE FP7 project test sites but also in future studies. First is the issue of terminology, and therefore a critical review was presented. It is beyond doubt that as regards the definition of terms such as hazard, vulnerability, risk, and the similar, there is need to reach at a consensus with the aim to avoid misunderstandings and for reasons of standardization.

A crucial finding is that at present the main effort regards the tsunami vulnerability of buildings. Very little has been done for other types of vulnerabilities, such as the vulnerability of communities in general, the vulnerability of humans, the vulnerability of properties indoor and outdoor and so on. However, some efforts have been made to assess the role that environmental factors play in the tsunami vulnerability. Such factors include coastal forests and other vegetation and coastal topography. From another point of view, the concept of time-dependent tsunami vulnerability practically has been neglected from the studies so far. Interesting findings were obtained about some of these issues from a tsunami questionnaire survey performed in Heraklion test site, Crete, in the frame of the EU FP7 ASTARTE Research Project (Papageorgiou et al., 2015). Similar results are expected from other ASTARTE test sites.

As regards the tsunami vulnerability of buildings, the development of computational methods has been rather neglected and only empirical methods, either qualitative or quantitative, were developed. The list of methods include mainly the PTVA model and its revised versions as well as the tsunami vulnerability assessment based on fragility functions (curves) and damage curves. The advantage of the initial PTVA model (Papathoma et al., 2003) and its later versions is that it provides estimations of the vulnerability levels for coastal zones in the lack of building damage data from real tsunami attacks. However, the inundation area is only roughly considered from historical data, while tsunami hydrodynamic parameters, such as flow depth or current velocity, are not taken into account. The introduction of fragility curves and damage curves, which are based on observational data collected after real tsunami events, improved the characterization of the expected damage level. Relevant tools were developed, such as the DamASCHE GIS tool (SCHEMA approach), to estimate the building damage level from an overlay of damage curves onto an inundation map. Both PTVA-3 model and DamASCHE GIS tool were very recently tested by several ASTARTE partners in other test sites and the results are expected with great interest to appear soon.

After the experience accumulated in the last 15 years or so but mainly in the post-2004 period, it was realized that the empirical methods in use for the assessment of the building vulnerability to tsunamis suffer from a variety of errors and uncertainties and, therefore, there is need for drastic improvement. In relevance to this realization, Tarbotton et al. (2015) suggested a series of recommendations for future work.

6.5 TSUNAMI RISK ASSESSMENT IN THE EM REGION

Globally the methods for tsunami risk assessment are less developed with respect to the methods for hazard and vulnerability assessment. This is well understood by simply looking in formula (6.2). This formula implies that methods for the hazard and vulnerability assessment should be first developed given that the risk is considered as a convolution of these two main parameters plus the value exposed to hazard. In the previous sections of this chapter it was unavoidable to refer to the international developments for the tsunami risk assessment. Therefore, here only some efforts made in this direction in the EM region are reviewed.

A very first attempt to qualitatively evaluate tsunami risk along the coasts of the Santorini Island due to future volcanic eruptions was published by Fritzalas and Papadopoulos (1988). After the paper by Papadopoulos and Dermentzopoulos (1998) regarding tsunami vulnerability and risk assessment (Section 6.4.1), most of the effort was focused on the risk assessment as a problem closely related to the vulnerability assessment, as it is expressed by the two main models developed, that is the PTVA model and its revised versions and the DamASCHE GIS tool (SCHEMA approach). In earlier times, a discussion on the assessment of tsunami size and implications for urban hazard planning in Greece was qualitatively discussed by Dominey-Howes (1998). As a follow up, the same author contributed to the issues of tsunami risk assessment and disaster management in Greece based on historical and geological records of tsunamis (Dominey-Howes, 2002).

An interesting contribution was the one regarding tsunami risk assessment for Malta Island (Camilleri, 2006). This author started from basic tsunami physics and sources in the Mediterranean, proceeded to tsunami vulnerability assessment based on the PTVA model and reached at calculating gross annual damage rate for a single tsunami event of 4–7-m high and mean return period between 600 years and 1500 years.

Tsunami risk studies were performed in the areas of western Mediterranean, SW Iberia, and Morocco by several authors in the last years. Birkmann et al. (2010) obtained qualitative data through, for example, focused group discussions and then developed hazard and vulnerability maps for the coast of Cadiz. They supported that their findings yield important information that could assist with the systematic improvement of societal response capacities of cities and their inhabitants to potential tsunami risks. Baptista et al. (2011) elaborated tsunami flooding maps for Lisbon downtown and the Tagus estuary. Such maps are of particular value for organizing evacuation maps and emergency plans. Renou et al. (2011) presented a generic methodology developed to produce tsunami-building vulnerability and impact maps. In a companion paper, Atillah et al. (2011) focused on the application of a GIS-based method to assess building vulnerability and damage in the event of a tsunami affecting the coastal area of Rabat and Salé, Morocco. Ribeiro et al. (2011) were based on tsunami simulations, produced detailed inundation maps and combined the results with the available information on the vulnerability of the local infrastructures (building types, roads and streets characteristics, priority buildings) in order to impose restrictions in the production of high-scale potential damage maps, escape routes and emergency routes maps. Such an approach was used to evaluate the specific risks associated with a potential occurrence of a tsunami event in the region of Setúbal (Portugal), which was one of the areas most seriously affected by the 1755 tsunami.

Other initiatives related to the tsunami risk assessment are the ones described by Sahal et al. (2013) for the French Riviera as well as by Omira et al. (2013) for El Jadida coast, Morocco. However, since these studies are also of interest to the tsunami risk mitigation they are further examined in Chapter 7.

In Istanbul, Marmara Sea, Hancilar (2012) produced inundation maps resulting from probabilistic tsunami hazard analysis for a 10% probability of exceedance in 50 years which was utilized in combination with the geo-coded inventories of building stock, lifeline systems and demographic data. In terms of social risk, it was estimated that there are about 32,000 inhabitants exposed to tsunami hazard.

A tsunami risk assessment study by following in the strict sense the formula (6.2) was presented by Grezio et al. (2012) for the Messina, Sicily. The hazard component was calculated by a new general modular Bayesian tool for PTHA, while several statistical data sets were used to estimate losses of persons and buildings. The time dependency of the human exposure to the tsunami hazard was also taken into account. Since the tsunami risk is a function of the runup levels along the coast, Grezio et al. (2012) defined risk as the area along the Messina coast where tsunami inundations may occur.

A very recent study regarding the historical record of tsunami impact in the entire EM region was presented very recently by Diakogianni et al. (2015). After organizing an inventory of the tsunami impact event by event, those authors mapped the spatial and temporal distributions of the several types of impact on human beings, buildings, vessels, and so on (Chapter 3). It was found that the main impact comes from extreme, earthquake-generated tsunamis, like the AD 365 and 1303 in Crete, 551 in Lebanon, and 1755 in Lisbon. However, high impact may also occur from lower magnitude events, such as the 1908 Messina and 1956 South Aegean earthquakes, which underlines the strong dependence of the impact on the community exposure. Another important finding is that the cumulative impact of relatively moderate or even small, local tsunamis, produced by earthquakes, landslides or volcanic activity, is quite important and that such distributed tsunami impact should not be neglected in actions undertaken for the tsunami risk mitigation.

CHAPTER 7

Tsunami Early Warning Systems and Risk Mitigation

7.1 INTRODUCTION

Tsunami waves in the European and Mediterranean (EM) region are generated mainly by submarine or coastal strong earthquakes but also by landslides and volcanic eruptions. A zonation of relative tsunami potential in the Mediterranean Sea and its connected seas is illustrated in Figure 2.1, while tsunami impact has been also reported in the North East Atlantic Ocean and in the Norwegian Sea along the western coasts of Norway (Figures 3.1 and 3.10). Observational data (Chapter 2) but also results of numerical modeling (Chapters 5 and 6) have indicated that the coastal zones of the EM region are threatened by near-field tsunamis since the tsunami sources are situated very close to coasts. Therefore, the first tsunami wave may arrive in the nearest coasts in time intervals ranging between about 5 min and 30 min. However, in case of large basin-wide tsunamis coastal zones at distances of 1 or more hours of tsunami travel times are also at risk. Historical examples in the Hellenic Arc are the large Late Bronze Age (LBA) Santorini tsunami and the Crete tsunamis of AD 365 and 1303 that propagated to remote places of the eastern Mediterranean basin. In SW Iberia, the 1755 large Lisbon tsunami propagated over long distances in the North East Atlantic Ocean. From the tsunami risk point of view it is of great importance that according to global tsunami statistics about 80% of victims due to tsunamis are caused within the first 1 h of tsunami propagation (Gusiakov, 2009). The need for undertaking in the EM region tsunami risk mitigation actions, including tsunami early warning systems, becomes obvious.

The establishment of organized tsunami early warning systems and of other actions aiming to mitigate tsunami risk in the EM region started after the Indian Ocean 2004 megatsunami. However, the need to develop such actions was stressed well before 2004 thanks mainly to the EU-supported tsunami research projects GITEC (1992–1995) and GITEC-TWO (1996–1998). Thanks to the positive impacts of those projects some early efforts were undertaken.

Tsunamis in the European-Mediterranean Region. http://dx.doi.org/10.1016/B978-0-12-420224-5.00007-7

After a short overview of the first efforts toward organizing early warning systems and other risk mitigation actions, this chapter focuses on the main efforts that were developed after 2004 at national, regional, and European–Mediterranean levels. The current status is examined and prospects for future improvements are discussed.

7.2 EARLY EFFORTS

Before 2004 the tsunami risk in the EM region was rather underestimated, yet risk ignorance was prevailing among decision makers as well as among the general public. Nearly everyone was aware about the "exotic" case of the megatsunami that was produced in prehistorical times by the LBA Thera (Santorini) volcano eruption and its supposed link with the Minoan civilization collapse. However, only few people were aware that tsunamis threaten the coastal zones of the EM region and that the tsunami risk increases with the increase of the community exposure along the coastal zones.

With the execution of the first well-organized, pan-European GITEC and GITEC-TWO tsunami research projects, the scientific knowledge about the generation mechanisms, the historical record, the hazard and risk assessment, and other aspects of the science of tsunamis in the EM region improved drastically. This scientific knowledge provided a good basis for someone to understand that the tsunami threat is an important issue and that there is urgent need for risk mitigation actions. Therefore, by the end of the project GITEC-TWO, an EU-supported special conference was held in Paris in May 1998 for the presentation of relevant results. The participation of civil protection representatives from some country members was quite promising for the future. The first step was done. Tsunami scientists and decision makers came together.

In the frame of GITEC-TWO a very first effort was made for the development of an experimental, local, telemetric tsunami warning system in the western segment of the Hellenic Arc, which is one of the most tsunamigenic zones in the EM region (Figure 2.1). The local system comprised a number of new seismic stations, in addition to the ones that the National Observatory of Athens was operating in the area in the frame of the Greek national seismograph system, two new tide gauges, and an acquisition unit installed at the premises of NOA in Athens for the data collection from the telemetric system (Papadopoulos, 2003c). Although this effort did not survive beyond the life of GITEC-TWO, it was revitalized after 2004 (Section 7.3). This experimental initiative, along with the positive impact of

Figure 7.1 ***The first tsunami alert unit in Stromboli.*** *(Photo by G.A. Papadopoulos, March 2003.)*

the 1998 workshop in Paris, contributed toward establishing a prevention culture as regards tsunami risk. In fact, the General Secretary for Civil Protection of Greece asked the country prefectures having coastal exposure to organize special tsunami emergency plans. Similar interest was expressed in other Mediterranean countries including Italy, France, Spain, and Turkey as well as Portugal.

At the same time, a simple, local alert system was installed in Stromboli comprising a tide gauge instrument to monitor the sea level changes and a sound-alerting device (Figure 7.1). This local system was installed after the tsunami event of December 30, 2002 that was generated by a volcanic landslide. The aim was to provide alerts in case the sea level exceeded a certain level, thus indicating possible tsunami arrival. However, a more complete local system for tsunami warning was developed later in Stromboli, which is described in more detail in a later section. A simple pilot, local alert device, similar to the one initially installed in Stromboli, was earlier installed in Augusta, eastern Sicily (Piscini et al., 1998), which historically has been hit by several tsunamis having their sources either in Sicily or in the western Hellenic Arc.

7.3 NEAMTWS: ASSOCIATED NATIONAL CENTERS - NEAMTIC

7.3.1 The North East Atlantic and Mediterranean Tsunami Warning and Mitigation System (NEAMTWS)

7.3.1.1 Establishment and Structure

In response to the devastating tsunami of December 26, 2004 in the Indian Ocean region (Chapter 1), the Intergovernmental Oceanographic Commission of UNESCO (IOC-UNESCO) received a mandate from the international community to coordinate the establishment of tsunami warning systems around the globe. This was realized during the course of several international and regional meetings, including the World Conference on Disaster Reduction (Kobe, Japan, January 18–22, 2005) and the Phuket Ministerial Meeting on Regional Cooperation on Tsunami Early Warning Arrangements (Phuket, Thailand, January 28–29, 2005). The IOC Assembly, during its twenty-third session (June 21–30, 2005), formally established the Intergovernmental Coordination Group (ICG) for the Tsunami Early Warning and Mitigation System in the North Eastern Atlantic, the Mediterranean and connected seas (ICG/NEAMTWS). In parallel, similar systems were developed in the Indian Ocean as well as in the Caribbean Sea. In the Pacific Ocean the system was already in place since 1965 as a response to the large, Pacific-wide tsunami caused by the May 22, 1960 big M 9.5 earthquake in Chile as well as to other tsunamis that hit the Pacific Ocean by the beginning of 1960s.

The annual assembly of the ICG/NEAMTWS, consisting of national delegations, makes the decisions and coordinates the NEAMTWS actions. The ICG convened its first session in Rome in November 2005. The ICG/NEAMTWS officers are elected by the country member's national delegations. Several Working Groups contribute on a voluntary basis for the production of new knowledge and know-how with the aim to support the system. Usually the WG contributions are combined with the works executed in the frame of other funded projects (e.g., EU FP6 TRANSFER and FP7 ASTARTE projects). Special Task Teams look after specific issues such as the architecture of the system, the operational works, the communications, and the exercises, thus providing to the ICG recommendations for the improvement of the system.

In the time period between two successive sessions of the ICG, the Steering Committee (SC) is the body that has been charged to promote and put in practice the decisions taken by the ICG. Also, the SC prepares proposals and suggestions for discussion in the next ICG by taking into account the progress achieved or problems noted in NEAMTWS as well as the recommendations received by the WG's and Task Teams. The schedule of progress in the NEAMTWS is described in, and organized around, periodically updated documents such as the Implementation Plan, the Operations Users Guide, and the Development Plan.

7.3.1.2 Infrastructures

From the point of view of the instrumental, communication, and database infrastructures, the NEAMTWS is supported mainly, but not exclusively, by the national centers that participate at the system. Networks of seismographs, including strong-motion instruments, and of coastal tide gauges, are the main recording systems, which are offered. The geographical coverage has been gradually increasing after 2004 and currently it is satisfactory. However, there are still several gaps, for example, in the North Africa region (Schindelé et al., 2008). GNSS networks are also available although they have not been integrated for operational purposes up to now. The intensive research that has been undertaken, for example, in the ongoing project ASTARTE, along with the increase of the number of stations in the frame of other projects (e.g., national, bilateral), are actions that promise for the integration in the not far future of the GNSS networks in the operational part of the NEAMTWS.

The cooperation of various national centers with international organizations as well as international initiatives of IOC itself act as a multiplier of the infrastructures, which are available for the NEAMTWS. For example, a long number of seismograph stations operated by the Comprehensive Nuclear-Test-Ban Treaty (CTBT) Organization, are available. The Global Sea Level Observing System (GLOSS) is an international program conducted under the auspices of the Joint Technical Commission for Oceanography and Marine Meteorology (JCOMM) of the World Meteorological Organization (WMO) and the Intergovernmental Oceanographic Commission (IOC). GLOSS aims at the establishment of high quality global and regional sea-level networks for application to climate, oceanographic, and coastal sea-level research. GLOSS tide gauge station records are also available for the NEAMTWS tsunami operations.

Of particular importance is the cooperation of several national tsunami centers with the tsunami program of the Joint Research Center of the European Commission in Ispra, Italy. In fact, the infrastructure support that the JRC provides includes instrumental sensors, for example, tide gauges, as well as software immediately relevant to tsunami warning operations. As analyzed in a next section, the JRC Tsunami Programme is an important component of the overall effort for the operation of early tsunami warning systems in the NEAM region.

So far, no sea bottom tsunami sensors have been integrated in the NEAMTWS. However, although discussions have been made as regards the possible installation and operational integration of such sensors in the NEAMTWS, it has been realized that some critical issues should be carefully examined before making decisions in this direction. These issues are examined later. Communication infrastructures consists of all the communication types available, including phone and fax lines, sms and email communications, and GTS. Of particular importance are the several databases that support the NEAM Tsunami Warning System, such as databases of presimulated tsunami scenarios as well as catalogs and relevant inventories of past earthquake and tsunamis, including databases of faults, seismogenic zones, volcanic fields, and landslide bodies. The presimulated tsunami scenarios are of crucial operational importance and, therefore, the methodologies used for their production are discussed further later.

7.3.1.3 Exercises and Training

The execution of exercises and drills, communication tests, and training actions is an important component for the NEAMTWS. The importance comes out from the general need that tsunami early warning systems must maintain a high level of readiness. In particular, in the NEAM region no operational experience with real tsunami events exist since no tsunami occurred after the summer of 2012 when the first national centers became operational. Therefore, one of the regular activities incorporated in the program of NEAMTWS includes periodic communication test exercises with the involvement of all centers and civil protection authorities of the NEAMTWS country members, of IOC itself, and other relevant message recipients.

The performance of tsunami scenario-based exercises is also a major activity toward maintaining readiness and executing training within the NEAMTWS programme. After the Tsunami Exercise NEAMWave12 that was performed in 2012, a major event was the Tsunami Exercise NEAMWave14, which took place on October 29 and 30, 2014 based on four different tsunami scenarios with a national center being charged for a respective scenario: North East Atlantic (IPMA, Portugal), Western Mediterranean (CENALT, France), Eastern Mediterranean (NOA, Greece), and Black Sea (KOERI, Turkey). The Tsunami Exercise NEAMWave14, which was supported by the IOC-UNESCO as well as by the European Commission, was formed in three phases.

Phase A was the simulation of the first step of the early warning process with the production and dissemination of tsunami alert messages to relevant subscribers. Phase B was open to member states by invitation and it was foreseen that different types of exercise could be included, such as an orientation exercise, a drill, a table-top exercise, or a functional exercise. Twenty countries participated in Phase B. Phase C of NEAMWave14 was only performed for the Eastern Mediterranean scenario prepared by the tsunami center of NOA. The performance was at international level on October 30, 2014 in order to simulate a real situation and tsunami-related emergency. The tsunami center distributed the initial tsunami messages concerning the earthquake and the tsunami. The Emergency Response Coordination Center (ERCC) of the European Commission got in contact with several civil protection authorities in potentially affected countries collecting enquires for assistance. The collected requests for international assistance and offers were forwarded by the Union Civil Protection Mechanism via ERCC.

Details about the preparation, execution and evaluation of the NEAMWave14 can be found in the very nice document "Tsunami Exercise NEAMWave14: From Early Warning to Early Response," published by UNESCO (2015) with the support of the European Commission. In the Conclusions section of that document it is stated that the key operational chain related to tsunami warning begins with data collection and ends with saving lives. Further, it was evaluated that NEAMWave14 provided a unique opportunity to all stakeholders to test existing operational capabilities of this chain system and to identify the areas of improvement, both in technical and operational aspects of the system.

During 2011, the "EU POSEIDON 2011" tsunami disaster simulation exercise was performed in Crete Island, Greece, as a project supported by the DG-ECHO of EU. The scenario, that was elaborated by local institutions, was a large earthquake on the Hellenic Arc, which would not only cause substantial structural damage to the island, but would also trigger a large tsunami that would hit most coastal areas of Crete. The scenario was planned so as to overwhelm local response capabilities and involved the triggering of assistance from the EU Civil Protection Mechanism. The simulation included a table-top exercise, a command post exercise, and a full-scale exercise. More details can be found in the conference presentation by Karagiannis et al. (2015).

There is no doubt that "EU POSEIDON 2011" was a positive exercise for local and EU civil protection authorities. However, it is not clear how the simulated involvement of rescue teams and other civil protection bodies responded specifically to disaster elements caused by the simulated tsunami or to a disaster caused by the earthquake in general. The exercise very likely would conclude even more positive results if the NEAMTWS mechanism was also involved.

7.3.2 NEAMTWS: Associated National Centers – Operational Status

Before 2004, France was the only European country having operational experience with the tsunami early warning thanks to the system developed in French Polynesia, which very likely is one of the oldest in the world along with the Japanese national warning system that started to operate in the 1950s.

The years that followed the 2004 devastating tsunami in the Indian Ocean signified a period that several countries of the Mediterranean, including Portugal, started to prepare their national tsunami warning centers and to participate at the North East Atlantic and Mediterranean Tsunami Warning and Mitigation System under the coordination of IOC-UNESCO (Figure 7.2).

7.3.2.1 Associated National Centers

The participation of country members is of vital importance for the functionality of the NEAMTWS. The country members are affiliated in the ICG/NEAMTWS by their designated Tsunami National Contacts (TNCs) and Tsunami Warning Focal Points (TWFPs). During the annual session of the ICG the country members are represented by the respective national delegations. However, the corner-stone for the operational part of the NEAMTWS are the National Tsunami Warning Centers (NTWCs) that have been accepted to provide tsunami services within the frame of

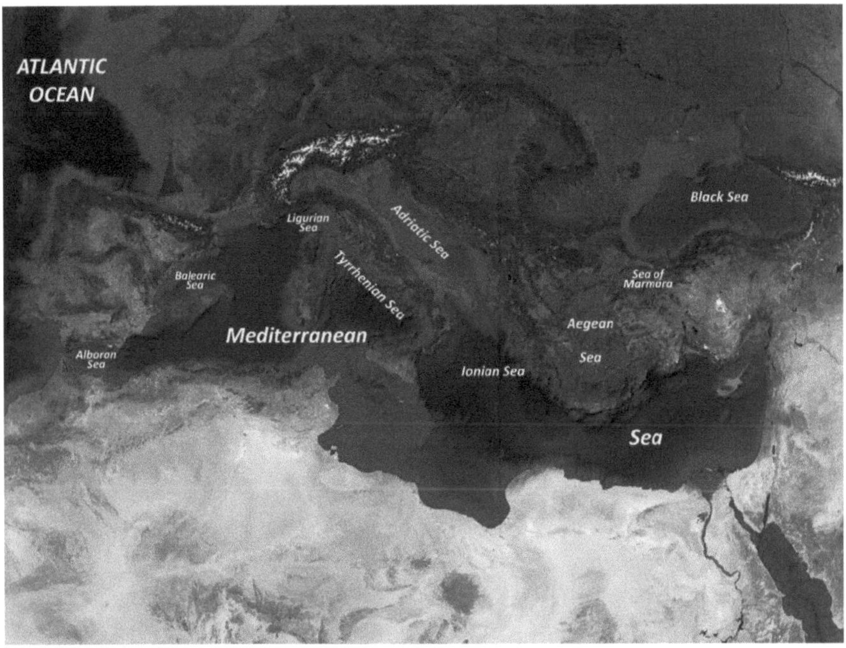

Figure 7.2 ***The North East Atlantic Ocean and Mediterranean Sea region.***

NEAMTWS. Since summer 2012, the national tsunami warning centers of France (CENALT, CEntre National d'ALerte aux Tsunamis), Greece (HL-NTWC, established at NOA, Figure 7.3), and Turkey (KOERI) started to provide such services. During 2014, the CAT (Centro di Allerta Tsunami, the Italian Tsunami Warning Center) established in INGV was added as a center providing tsunami services.

The tsunami centers of CENALT, INGV, KOERI, and NOA are formally still named "Candidate Tsunami Service Providers (CTSPs)" since the procedure of their accreditation has not yet been completed, but they are expected to be completed soon. In practice, those centers already act operationally at both the national level and at the NEAM level. Also, Germany has offered to provide backup for data collection and processing through the GFZ technological infrastructures. In Portugal, after an initial phase that included the development of a Prototype for the Portuguese Tsunami Warning System (PtTWS; Annunziato et al., 2009; Omira et al., 2009) getting support from the FP6 NEAREST project, the Portuguese national tsunami center has been established in IPMA in Lisbon. It is expected that in the near future IPMA will join the group of national centers providing tsunami services within the frame of NEAMTWS.

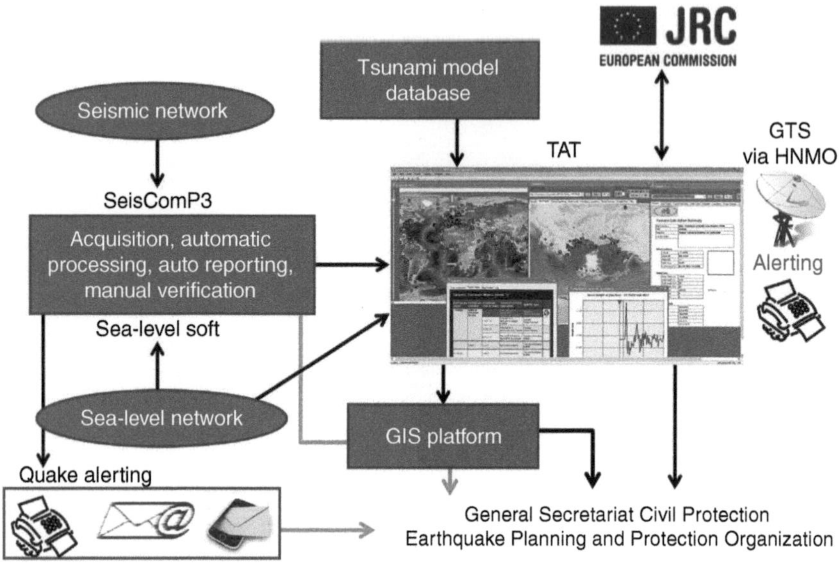

Figure 7.3 ***Schematic diagram illustrating the operational function of the Hellenic National Tsunami Warning Center at NOA, Greece.***

It is understood that in their national roles the tsunami centers act according to the respective national mandates and instructions, which are adjusted to the national administrative needs that may be different from the ones at the NEAMTWS level. However, from the technical point of view the standards that the national centers follow at national level are the same with those at the level of NEAMTWS. In fact, all the centers that provide tsunami services have in common some standard characteristics decided by the ICG/NEAMTWS, such as the operation on 24/7 basis of national seismograph systems and tide gauge networks as well as expertise in seismic monitoring and tsunami science.

7.3.2.2 Operational Status of NEAMTWS

In their operations, the centers that provide tsunami services within the frame of NEAMTWS follow some common, basic steps. Namely, as soon as an earthquake of minimum magnitude 5.5 has been generated under the sea or close to the coastal zones, the tsunami center is responsible to collect and record, process and analyze earthquake data for the rapid initial assessment as a basis for the alert system. The basic seismic information which should be determined as soon as possible after an earthquake is detected are the main focal parameters of the seismic event, that is the earthquake origin time, epicenter, focal depth, and magnitude.

The next step is to follow a decision making process in accordance with the operational plan approved by the ICG/NEAMTWS with the aim to formulate tsunami messages. Since it is not known whether a tsunami has been generated or not immediately after the earthquake is detected, a tool called the "decision matrix" is used to decide the type of tsunami message that should be formulated and disseminated. A "decision matrix" is a set of empirical rules regarding the possibility for a tsunami to have been generated or not according to the location of the earthquake epicenter as well as the earthquake focal depth and magnitude. After long technical discussions within the working groups that support the ICG/NEAMTWS, it was decided that two similar in their structure but different in the numerical values "decision matrices" are appropriate to use, one for the North East Atlantic region and another for the Mediterranean basin.

In Tables 7.1 and 7.2, one may see that the tsunami severity predicted by the "decision matrices" scales with the earthquake magnitude. As a consequence, three levels of tsunami messages could be produced with the increase of the wave severity: *tsunami information*, *tsunami advisory*, and *tsunami watch*. At the same time, the magnitude range determines also the maximum distance at which tsunami impact is likely to be caused in the coasts. From this point of view local, regional, and basin-wide tsunami message types are considered. *Local distance* means no more than 100 km. *Regional distance* has been decided to be up to 400 km for the Mediterranean region and up to 1000 km for the North East Atlantic Ocean. *Basin-wide distance* is the spatial field of possible tsunami impact beyond 400 and 1000 km for the Mediterranean and North East Atlantic, respectively. The assessment of the possible tsunami wave severity is summarized in Table 7.3.

However, the interest is not restricted only in making decisions if a tsunami has been potentially generated and to assess its overall severity. On the contrary, it is of alerting importance to compute also the arrival time of the tsunami in a number of coastal forecast points that have been determined officially by the NEAMTWS country members. Therefore, besides earthquake focal parameters, a standardized tsunami alerting message informs the recipients about the type of the tsunami message, using combinations between the various severity types (information, advisory, watch) and the types of possible impact distance (local, regional, basin-wide), as well as the calculated arrival time in a number of official forecast points. The tsunami message should be sent as soon as possible to a list of relevant subscribers, which include national civil protection authorities of country

Table 7.1 Decision matrix for the North East Atlantic Ocean, proposed by the ICG/NEAMTWS in November 2010

				Tsunami message type		
Focal depth (km)	**Epicenter location**	M_w	**Tsunami potential**	**Local**	**Regional**	**Basin**
<100	Offshore or close to the coast (≤40 km inland)	>5.5 and ≤7.0	Weak potential for local tsunami	Advisory	Information	Information
		>7.0 and ≤7.5	Potential for a destructive local tsunami (<100 km)	Watch	Advisory	Information
	Offshore or close to the coast (≤100 km inland)	>7.5 and ≤7.9	Potential for a destructive regional tsunami (<1000 km)	Watch	Watch	Advisory
		>7.9	Potential for a destructive basin-wide tsunami	Watch	Watch	Watch
≥100	Offshore or inland ≤100 km	>5.5	Nil	Information	Information	Information

Table 7.2 Decision matrix for the Mediterranean basin, proposed by the ICG/NEAMTWS in November 2010

Focal depth (km)	Epicenter location	M_w	Tsunami potential	Tsunami message type: Local	Regional	Basin
<100	Offshore or close to the coast (≤40 km inland)	>5.5 and ≤6.0	Weak potential for local tsunami	Advisory	Information	Information
		>6.0 and ≤6.5	Potential for a destructive local tsunami (<100 km)	Watch	Advisory	Information
	Offshore or close to the coast (≤100 km inland)	>6.5 and ≤7.0	Potential for a destructive regional tsunami (<400 km)	Watch	Watch	Advisory
		>7.0	Potential for a destructive basin-wide tsunami	Watch	Watch	Watch
≥100	Offshore or inland ≤100 km	>5.5	Nil	Information	Information	Information

Table 7.3 Tsunami message type and respective tsunami wave severity

Message type	Tsunami wave	Effects on the coast
Tsunami watch	Tsunami wave height greater than 0.5 m and/or tsunami runup greater than 1 m	Coastal inundation, potentially including all the phenomena described as potential coastal effects corresponding to tsunami advisory
Tsunami advisory	Tsunami wave height less than 0.5 m but larger than 20 cm and/or tsunami runup less than 1 m	Currents, bore, recession, damage in harbors, small inundation on beaches
Tsunami information		No tsunami threat

members, other NEAMTWS centers and the IOC itself, as well as the ERCC of the European Commission.

The next step in the alerting procedure is to collect, record, process, and analyze sea-level data for confirming and monitoring the tsunami or for canceling elements of the alert message. If a tsunami occurrence has been confirmed by the sea-level data then the tsunami message should be updated with the relevant information and resent as soon as possible. If needed, this procedure may repeat more than once. When there is no need to repeat it further a cancellation message is issued and disseminated. If, however, no tsunami generation is confirmed, then a tsunami cancellation message is issued and disseminated.

In their operations the tsunami centers should be *rapid*, by providing warnings as soon as possible after a potential tsunami generation; *accurate* by issuing warnings for all potentially damaging or destructive tsunamis while minimizing the false warning rate; *reliable*, by making sure they operate continuously, and that their messages are sent and received promptly and understood by the users of the system. It should be mentioned that the previous procedures, which are followed by the national centers providing tsunami services in the frame of NEAMTWS apply in cases of tsunamis that could be generated by earthquakes. NEAMTWS, however, has no mechanism to warn for tsunamis that could be generated by aseismic landslides or by volcanic activity. This is exactly the operational principle of the tsunami warning systems operating under the coordination of respective ICGs in the Pacific Ocean, the Indian Ocean, and the Caribbean Sea. Of interest is also that since 2010 an effort has been undertaken to harmonize and standardize the operations of the four systems operating under the umbrella of IOC-UNESCO. Since then significant progress has been made by the Inter-ICG Task Teams on tsunami watch operations and on disaster management and preparedness working within the mandates given by the Working Group on Tsunamis and Other Hazards Related to Sea-Level Warning and Mitigation Systems (TOWS-WG) established by IOC.

It is noteworthy that no responsibility areas have been determined for each one of the centers that operate in the frame of NEAMTWS. Therefore, sending and receiving tsunami messages is performed on a subscription basis. Nevertheless, the practice followed says that centers, for example, NOA (Greece) and KOERI (Turkey), being established in the area of eastern Mediterranean, do not produce tsunami messages for the North East Atlantic region. On the other hand, CENALT do not produce tsunami messages for the eastern Mediterranean and so forth.

7.3.2.3 Other National Tsunami Centers and Services in the Mediterranean

Other national tsunami centers are also of interest even though they have not yet integrated into the NEAMTWS. For example, on the coasts of Israel the tsunami risk is not negligible since in the past not only local tsunami events but also remote ones, having their sources in the Hellenic Arc, have been reported (Fokaefs and Papadopoulos, 2007; Salamon et al., 2009). In view of this fact, national tsunami emergency plans are elaborated by the relevant national Israeli authorities, while the initiative is supported by a well-organized telemetric seismograph system and other instrumental networks. In addition, the Israeli national delegation participates regularly and actively in the annual sessions of ICG/NEAMTWS as well as to the various WGs of the ICG, being closely associated with the IOC Tsunami Program for the NEAM region. There is an agreement for communication between the Israeli emergency management authority and the national tsunami warning center of Greece at NOA in cases of strong, potentially tsunamigenic earthquakes occurring along the Hellenic Arc. The aim is to collect and evaluate information about potential tsunami generation that may affect the coastal zones of Israel.

In Cyprus, a new telemetric seismograph system, consisting of nine BB land stations, replacing the old system, was established a few years ago by the Geological Survey of Cyprus (GSC) situated in Nicosia. Also, two OBSs were installed offshore SW Cyprus sending signals to GSC as well as to NOA in Athens. In parallel, a committee regarding the tsunami hazard in Cyprus was established under the leadership of GSC. In addition, the Oceanography Center of the University of Cyprus (OC-UOC) upgraded the network of tide gauges thanks to its participation in several projects, including the project NEARTOWARN (2012–2013) supported by the EU DG-ECHO and coordinated by NOA, Greece. The link of Cyprus with the EU and the NEAMTWS initiatives strengthened with the organization in Nicosia of the annual meeting of NEARTOWARN (November 2012) and the eleventh Session of ICG/NEAMTWS in November 2014. In view of these achievements there are good reasons to believe that a national tsunami warning center could be established in Cyprus.

The participation of national delegations from countries like Egypt, Lebanon, and Syria in sessions at the ICG/NEAMTWS are encouraging for their closer integration with NEAMTWS in the future. In Alexandria, Egypt, a national workshop with the participation of international experts was organized by national scientific institutions and authorities and convened in May 2012. Prospects for the creation of a national tsunami warning system were examined.

At the other side of the Mediterranean basin, Morocco has made remarkable progress by participating in various EU and bilateral tsunami projects. Representatives of governmental, scientific, and emergency management institutions from the Maghreb countries gathered in Rabat (Morocco) on September 23–24, 2014 for a workshop on tsunami warning and emergency response in the region organized by local institutions jointly with IOC, NEAMTWS, and NEAMTIC. It is promising that Moroccan, Algerian, and Tunisian experts presented their efforts toward the creation of national tsunami warning centers and their plans to integrate more actively with NEAMTWS.

7.3.2.4 Western Black Sea Initiative

In the last years, a remarkable effort started for the coverage of western coast of Black Sea from the point of view of tsunami early warning. This initiative responds to the fact that although Black Sea is of low tsunami activity, the tsunami risk there should not be neglected as is indicated by the few strong tsunami events reported in the historical past (Papadopoulos et al., 2011). For the accomplishment of this goal, the Romanian National Institute for Earth Physics (NIEP) collaborated with the National Institute for Marine Geology and Geoecology (GeoEcoMar), the Institute of Oceanology and the Geological Institute, the last two belonging to the Bulgarian Academy of Science (Ionescu et al., 2014). The main purpose of the project was the implementation of an integrated early warning system accompanied by a common decision-support tool and enhancement of regional technical capability for the adequate detection, assessment, forecasting, and rapid notification of natural marine geohazards for the Romanian–Bulgarian Black Sea cross-border area. In this respect, NIEP has developed a coastal network for marine seismicity, by installing new seismic stations in the coastal area of the Black Sea, sea-level sensors, radar and pressure sensors, and GNSS stations at every site, providing tide gauges and seismic data exchange with the Black Sea countries (Ionescu et al., 2014). At the same time, the Tsunami Analysis Tool (TAT) software, for inundation modeling, along with its RedPhone application, were also installed at the National Data Center in Magurele city, and also at Dobrogea Seismic Observatory in the city of Eforie Nord, close to the Romanian Black Sea shore. Although the NIEP effort is not integrated in the NEAMTWS yet, it would be realistic to expect its integration in the near future.

7.3.2.5 The UK National Tidal and Sea Level Facility

After violent storms occurring in the North Sea in 1953, which caused severe flooding in the Thames Estuary, the UK national network of sea-level gauges was established. The National Tidal and Sea Level Facility (NTSLF) project has been developed by the United Kingdom since 2002. It is based on a network of more than 44 tide gauges around the UK coast complemented by geodetic networks, to monitor vertical land movements. Other tide gauges are located around the British Dependent Territories of the South Atlantic and Gibraltar. This type of system can be useful in case of far-field tsunamis, like a repeat of the 1755 Lisbon large tsunami or even of the large prehistoric Storegga landslide-generated tsunami (Horsburgh et al., 2008; Hill et al., 2014). However, the system is not suitable for near-field tsunami sources although no such sources have been identified so far.

7.3.3 NEAMTIC: Objectives and Approaches

The operation of instrumental tsunami early warning systems must go hand in hand with awareness and preparedness of the general public as well as with the training of the staff working with warning systems and with relevant decision-making authorities. In this way an end-to-end system for early warning and crisis management could be really useful for saving lives and risk mitigation. In view of these realizations, the IOC-UNESCO decided to establish the NEAM Tsunami Information Center (TIC) aiming to support the operational activities of the NEAMTWS. This initiative is similar to information centers supporting other systems, such the International TIC (ITIC) and the IOTIC that support the tsunami warning systems in the Pacific Ocean and the Indian Ocean, respectively. As part of this process, the NEAMTIC was established to provide information on warning systems, risks and good practices in respect of tsunamis and other sea-level related hazards for civil protection agencies, disaster management organizations, decision makers, schools, industries in the coastal zone, and the general public.

One of the objectives of NEAMTIC is to ensure people in endangered coastlines are prepared and know what to do in case a tsunami warning has been issued by the national authorities. Also, to better educate communities about tsunami threats, and commit to continue the effort that has been started by NEAMTIC. More specifically, for the realization of its objectives, the NEAMTIC project objectives are aiming to:

- Make citizens, especially youth, aware of risks of floods from the sea in coastal areas, such as tsunamis, storm surges and strong swells.
- Acquire knowledge on and practicing safe behavior.
- Identify, share and disseminate good practices in plans, methods, and procedures to strengthen preparedness for sea-level related hazards, including mitigation through integrated coastal zone management approaches.
- Fostering linkages between the European Commission and the IOC on intergovernmental and transnational actions to develop the NEAMTWS.

To fulfill its objectives, NEAMTIC has undertaken a series of actions:

1. Development and distribution of multilingual education, awareness and preparedness materials, like booklets, posters and e-learning tools.
2. Identification and exchange of good practices and assist civil protection authorities in the establishment of national and regional warning systems on coastal inundation, and the reduction of risks from tsunamis and other sea-level related hazards through comprehensive mitigation programs, including integrated coastal zone management.
3. Provision of information on warning systems for tsunamis and other sea-level related hazards to civil protection authorities.

Several of the NEAMTIC activities were supported by the European Commission, for example, through the DG-ECHO programs, and were materialized thanks to the dedicated commitment of IOC staff. More details about the NEAMTIC activities as well as downloadable materials and information can be found in the web site http://neamtic.ioc-unesco.org/.

7.4 THE JRC TSUNAMI PROGRAMME

The Institute for the Protection and Security of the Citizen of JRC (Joint Research Center, European Commission) provides research-based systems oriented support to EU policies so as to protect the citizen against risks of various origin, including earthquakes, tsunamis, tropical cyclones, and floods. In this view, after the 2004 Indian Ocean megatsunami the JRC developed a tsunami program aiming to contribute in the tsunami risk mitigation. In the subsequent sections, the main issues related to the JRC Tsunami Programme are reviewed. Then, the contribution of this program in the NEAM region is analyzed.

7.4.1 The Global Disasters Alerts and Coordination System

The Global Disasters Alerts and Coordination System (GDACS) is a joint United Nations (OCHA) and European Commission (ECHO, JRC) system concerning several types of disasters. However, GDACS does not perform physical observations, such as deep sea observations or seismic records in case of earthquake and tsunami occurrences. Instead, it picks up such information from seismological organizations through web protocols and performs additional processing, like overlaying information with population density in order to estimate the potential impact. GDACS aims at improving the post-disaster information flow, including fast alerts, updated news, satellite maps, and needs and relief related information. In case of an earthquake, for example, when a new event is detected by seismological agencies (e.g., USGS, EMSC), an evaluation is performed to estimate the importance of the event from humanitarian point of view. If the event is relevant, the system automatically sends out alerts (email, SMSs, fax) to the registered users, which amounts to more than 23,000 by July 2015. The information is published on the GDACS web site in real time. Of specific value for tsunami early warning systems is the JRC Tsunami Assessment Tool, which became operational since November 2005.

7.4.1.1 JRC Tsunami Assessment System

The JRC Tsunami Assessment System (Annunziato, 2007a,b) is a complex computer arrangement whose objective is to predict the behavior of a tsunami when minimal parameters are known, that is under the condition that a strong earthquake, say of $M \geq 6.5$, is firstly measured. Therefore, knowing the epicenter and the magnitude of the earthquake, the program estimates the fault characteristics, the tsunami generation and displacement, and the identification of the location on the coast, which will be most likely affected. The JRC Tsunami Assessment System integrates several components that are needed in order to fully evaluate an earthquake-generated tsunami. The program uses empirical relationships and evaluates the fault length, direction and amplitude of displacement, the last influencing the initial water displacement. Then, the program initializes the calculation space, performs calculation of the travel time propagation and tsunami height at coastal forecasting points, verifies at each step if there are locations reached by the wave, and updates the visualization and animation files.

In order to cut the running time of the online calculations, a database of presimulated tsunami scenarios has been generated and as soon as the main focal parameters of an earthquake has been determined the program finds out the most appropriate scenario solution. A global database has been created, which for the Mediterranean area contains about 8000 scenarios. This database is used to have a first rapid estimation of the event; at the same time, however, the system launches one calculation for each new set of estimated earthquake parameters.

7.4.2 The Role of JRC in the Tsunami Risk Mitigation in the EM Region

In the frame of the GDACS and the related tsunami program, the JRC in close collaboration with the DG-ECHO, plays an important, positive role in the efforts for tsunami risk mitigation in the European–Mediterranean region. The relevant activities include software development for early warning purposes, organization of workshops for know-how exchange and coordination of actions, installation and development of new tide gauge stations, training projects and MoUs for collaboration with several national tsunami centers and civil protection authorities.

The JRC computer TAT, including the database, which for the Mediterranean area contains presimulated tsunami scenarios, has been freely available to several national tsunami centers, such as NOA (Greece), KOERI (Turkey), IPMA (Portugal), IGN (Spain), CNRST (Morocco), and NIEP (Romania). Currently some of these centers use TAT regularly for their tsunami operations in the frame of national activities and of NEAMTWS. Given that some of the tsunami centers operating in the frame of NEAMTWS have prepared or are in the process of preparation of their own scenario databases, TAT is an alternative or an addition to other databases. In any case, TAT has been proven a valuable tool in the last years, particularly since summer 2012 when the centers that support NEAMTWS gradually become operational. A discussion regarding pros and cons of the TAT scenario database with respect to other strategies available can be found later in a relevant section.

Installation of new tide gauges is an important activity offered by JRC in collaboration with national tsunami centers. For example, in collaboration with NOA (Greece) three tide gauges were installed in critical coastal points of the western segment of the Hellenic Arc, thus covering a gap existing in the sea level monitoring network in that segment. Similar activity was developed between JRC and other national tsunami centers in the region covered by the NEAMTWS. A novel type of low cost mareograph has also been developed by JRC and offered to the NEAMTWS countries for testing and evaluation; the relevant dissemination campaign is under way.

The low rate of tsunami occurrence in the NEAM region does not provide frequent opportunities for the staff of the tsunami centers to train themselves in real conditions for tsunami warning. This is a shortcoming for maintaining the staff in full readiness when it would be really needed. However, the operational procedures followed by the national centers providing tsunami watch services in the NEAMTWS frame as well as the exercises performed within NEAMTWS maintain readiness at a satisfactory level. Another opportunity was provided during 2013–2014 by the Global Tsunami Informal Monitoring Service (GTIMS-1), a training project supported by the JRC with the participation of national tsunami centers of the NEAMTWS region. At the time of writing, the GTIMS-2 was just launched for the period 2015–2016 with the participation of CNRST (Morocco), IPMA (Portugal), KOERI (Turkey), NIEP (Romania), and NOA (Greece). Apart from keeping the national centers' staff trained and in good readiness, the GTIMS-2 project also aims to support the Emergency Response Coordination Center (ERCC) of the European Commission.

The core idea for both GTIMS-1 and GTIMS-2 is that in case of a large ($M \geq 7$) earthquake anywhere on the globe, the staff of the participating national centers act immediately for the collection of information regarding the earthquake focal parameters as well as the possible impact of the earthquake and of the tsunami, if any. Relevant reports are organized and submitted as soon as possible. About 15 such earthquake events are expected annually as an average, which provides a good basis for regular training of the staff in the tsunami centers of the EM region.

7.5 OTHER EARLY WARNING SYSTEMS AND OPERATIONAL TOOLS

7.5.1 GITEWS: A German–Indonesian Project

In Germany, several institutes under the leadership of GFZ got important experience in the science and technology of tsunami early warning systems through the organizational effort of the German Indonesian Tsunami Early Warning System (GITEWS) that eventually was ready for operations by the end of March 2011 (for more details, see Rudloff et al. (2009), vol. 10 and 11). The system included also training and educational activities.

7.5.2 The NEARTOWARN Tsunami Travel Time Mapping Tool

The time needed for a tsunami to travel at a particular coastal point obviously depends on the distance from the source to the point as well as on the bathymetric features of the tsunami travel path. The knowledge of tsunami travel times is useful in the stage of preparing plans for management crisis, immediately after a potentially tsunamigenic earthquake generation as well as for hazard assessment and other tsunami studies in general. From this point of view, the availability of a tool for mapping tsunami travel times in an easy and practical way certainly is useful.

In the frame of the NEARTOWARN tsunami project (2012–2013) supported by the DG-ECHO, European Commission, and coordinated by the author of this book, among other activities a Tsunami Travel Time Mapping Tool (TTTMT) was developed, which is useful to decision makers and civil protection authorities as well as to tsunami scientists in general. The computer tool was developed by the NEARTOWARN partner IH Cantabria, University of Cantabria, Spain (Zorrilla Aniel-Quiroga and González, 2012, 2013). This computer tool allows to online retrieve descriptions and locations of seismic and nonseismic tsunami sources of the NEAM region from the relevant databases created by many contributors for the FP6 TRANSFER project. With the tool it is also possible to calculate inverse tsunami travel times and map travel time contours for NEAM countries and their main islands (Figure 7.4a–c) and also the travel times from each tsunami source in the NEAM region to each forecast point determined for the needs of NEAMTWS (Figure 7.5).

The TTTMT viewer can be found on the Internet, on the web site http://www.neartowarn.ihcantabria.com/ (use Google Chrome or Firefox, not IE). Travel times have been calculated using the MIRONE system (Luis, 2007) modified by IH Cantabria (Olabarrieta et al., 2011). MIRONE computes the tsunami travel time according to formula (7.1), which provides

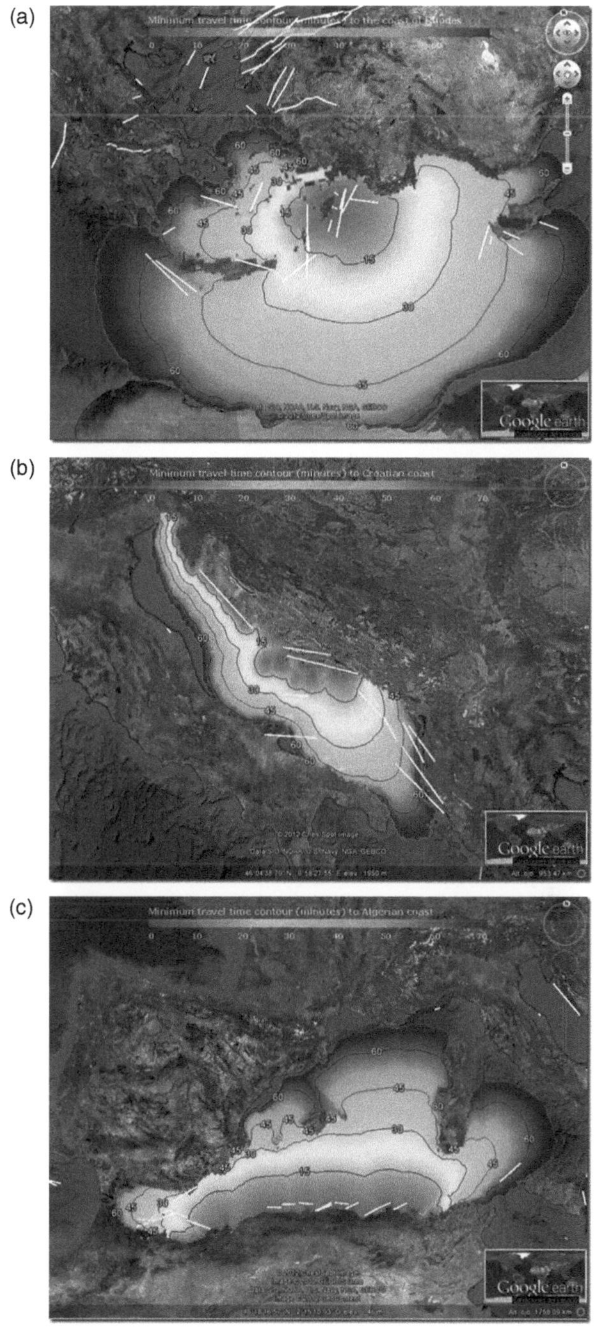

Figure 7.4 Isochron contours (in min) of inverse tsunami travel times calculated by Zorrilla Aniel-Quiroga and González, (2012) for (a) the city of Rhodes, situated at the northernmost tip of the island of Rhodes (b) the eastern-central Adriatic Sea; and (c) the coast of North Algeria. White lines represent fault lineaments.

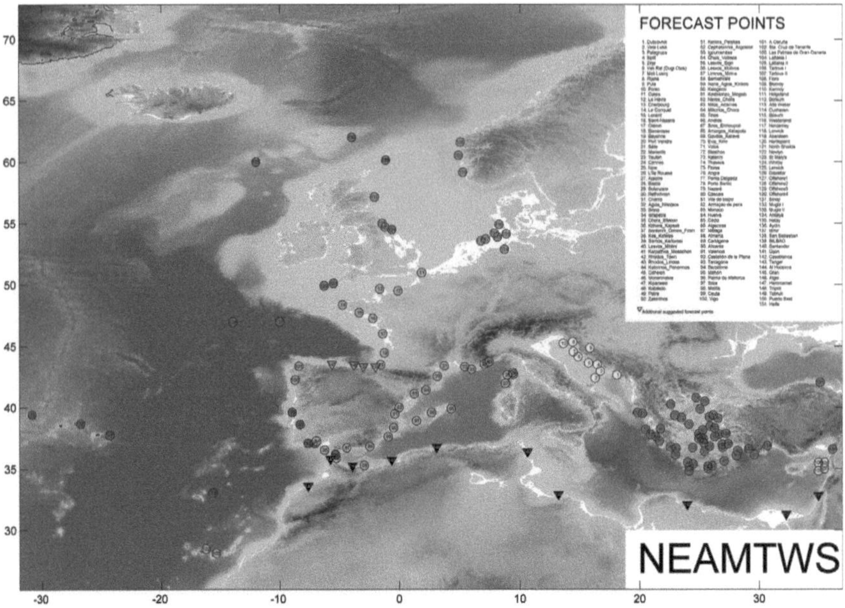

Figure 7.5 ***The forecast points in the NEAM region as determined by the country members of IOC represented in the ICG/NEAMTWS.*** Additional forecast points were installed along North Africa (gray triangles) by Zorrilla Aniel-Quiroga and González (2012), in consultation with the author of this book, for the needs of the TTTMT viewer tool.

a first approximation of the tsunami velocity, v, as a function of water depth, h, where g is the acceleration due to gravity:

$$v = \sqrt{gh} \tag{7.1}$$

In order to calculate the travel times, regular bathymetry grids have been constructed using the GEBCO database (Δx = 30 s mesh). The calculated travel times have been validated comparing them to results of numerical propagations using the C3 (COMCOT–CANTABRIA–CLAW) model that was developed and validated in the EU FP6 TRANSFER Project. The diffraction effect has not been relevant, thus, the ray propagation theory is a good approximation for the tsunami travel times calculation. More details on the computation techniques as well as about the access and use of the TTTMT viewer can be found in the reports of Zorrilla Aniel-Quiroga and González (2012, 2013).

7.6 LOCAL WARNING SYSTEMS FOR NEAR-FIELD TSUNAMIS

Apart from the national tsunami centers and the tsunami early warning systems and services of NEAMTWS, NEAMTIC, and JRC Tsunami Program, several local warning systems have been developed so far in various parts of the EM region. The need for tsunami early warning in the near-field (or local) domain has been internationally discussed not only after the Tohoku 2011 megatsunami but also before its occurrence (Schindelé, 1998; Murata et al., 2010). One definition for near-field tsunamis is that wave travel distances are of a similar order (of magnitude) to the earthquake rupture length (Lauterjung et al., 2010). Such physical preassumptions were adopted for the development of the German Indonesian Tsunami Early Warning System (GITEWS). It is expected that information of this type will be available 5–10 min after the event at best. More simply, I consider that in the EM region the term near-field should apply when the first tsunami arrival time, at a single forecast point or at a set of forecast points, is earlier than the first warning time supplied by national centers and NEAMWTS. This is further explained with the analysis of the Rhodes Island near-field tsunami warning system, which is the first case examined in this section.

7.6.1 REWSET: Rhodes Early Warning System for Earthquakes and Tsunamis

7.6.1.1 Background – Structure of the System

As already noted in Chapter 2, Rhodes Island (Greece) is situated in the easternmost segment of the Hellenic Arc, which is characterized by very high earthquake and tsunami activity as is indicated by geological, historical, and instrumental records (Papadopoulos et al., 2007a, 2012a). Historically, Rhodes was hit by large, earthquake-generated destructive tsunamis in AD 1303, 1481, 1609, and 1741. The largest, nontsunamigenic but destructive earthquake (M 7.2) in the last century occurred on April 25, 1957 (Papadopoulos, 2012a). Yet, a local aseismic tsunami was reported to cause damage in a coastal zone of Rhodes city on March 26, 2002. The permanent population of Rhodes is about 130,000, which at least doubles during summer peak season with a very dense people concentration along several coastal zones of the island. Rhodes, along with the town of Fethiye (SW Turkey), was the main test site for the FP6 TRANSFER tsunami research project (2006–2009).

One of the main activities of the NEARTOWARN project (2012–2013) supported by the DG-ECHO (European Commission) has been the development of the local and operational Rhodes Early Warning System for Earthquakes and Tsunamis (REWSET), which is composed by three main subsystems (Papadopoulos and Fokaefs, 2013).

1. Seismic alerting devices (SADs) network, consisting of eight ground motion sensors equipped also with light and sound alerting devices. Each SAD unit consists of two sensors, therefore four SAD units are

Figure 7.6 (a) One of the municipality buildings where SAD unit was installed in the frame of the REWSET (NEARTOWARN project) (photo by G.A. Papadopoulos, June 2012). (b) The master seismic alert sensor (lower part of the photo) of the SAD unit installed in the interior of the building shown in Figure 7.6a (photo by G.A. Papadopoulos, June 2012). Two sound and light alert devices connected with the sensor are shown at the upper part of the photo.

installed in four different localities of the island: one in the Civil Protection building, one in the Fire Brigade building, another two in other municipality buildings (Figure 7.6a,b).

2. Two radar-type (ultrasonic) tide gauges installed in the eastern coastal zone of the island (Figure 7.7), which was selected since research on the historical earthquake and tsunami activity has indicated that the most important, near-field tsunami sources are situated offshore to the east of Rhodes.
3. A geographic management system (GMS) developed and managed by the civil protection office. This is a web- and GIS-based application incorporating a variety of thematic maps and other information types useful for the organization, management, analysis, and spread of spatial information and data. The user, for example, disaster manager or citizen etc., can choose from a variety of thematic maps concerning human geography, geology, 3D surface (based on contours every 1 or 2 m), Cadastre (aerial photographs), land use, etc. After choosing a map, the user can add data and research results related to damages in buildings located in high risk zones (schools, hospitals, public services, hotels, churches, roads), in case of various natural hazards in Rhodes, such as earthquakes,

Figure 7.7 ***One of the ultrasonic (radar-type) tide gauges installed in eastern Rhodes in the frame of the REWSET (NEARTOWARN project).*** *(Photo by S. Liakopoulos, December 2013.)*

landslides, flooding/inundation, etc. Tsunamis are also included and they can be selected from a classification according to wave height. In each case several emergency elements are shown, like reception areas, directions to go there and more.

The REWSET became operational in July 2012 as regards the SAD network and in December 2013 as regards the tide gauges and the GMS. The SAD network has been set up to activate by strong earthquakes (magnitude about 6 or more) occurring at distances up to about 100 km from the city of Rhodes, thus providing immediate mobilization of the Rhodes Civil Protection authority and the fire brigade of the island. The seismic alerting signals are received at the installations of those authorities and also they are sent to selected persons via cell phone messages. The tide gauges transmit in

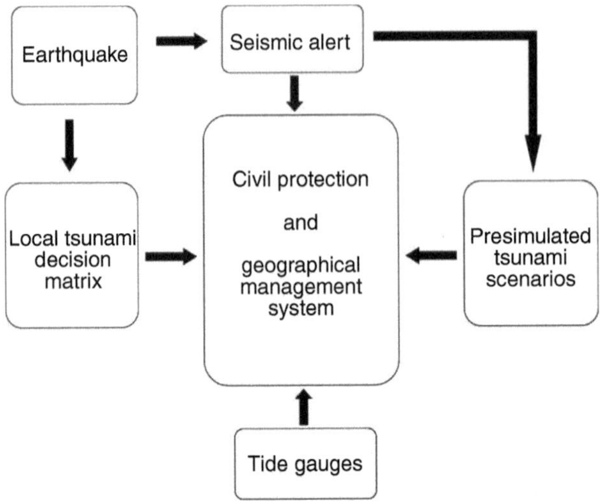

Figure 7.8 ***Flowchart of the REWSET (Papadopoulos and Fokaefs, 2012).***

real-time sea level data to the national tsunami center of NOA at Athens but also to the civil protection office of Rhodes Municipality, as well as to JRC at Ispra (Italy). In case of a crisis the GMS supports decisions to be made by the civil protection office. The flowchart of the REWSET is illustrated in Figure 7.8.

REWSET is not only an operational system for seismic and tsunami early warning in the near-field domain, but also a prototype one which certainly could be developed in other coastal areas of the Mediterranean and beyond.

7.6.1.2 The Concept of Warning in Near-Field

For designing and constructing REWSET as an operational near-field system, it was considered that the conditions for the operative efficiency of a tsunami early warning system strongly depends on the times needed for seismic signal communication and for people evacuation (Sasorova et al., 2008; see section 1.8.2). More analytically, the next relation should be realized:

$$t_{tr} > t_{ED} + t_{TM} + t_{TW} + t_{EV} \tag{7.2}$$

where t_{tr} is the first tsunami wave travel time from the source to the closest settlement and t_{ED}, t_{TM}, t_{TW}, and t_{EV} are times needed for the earthquake focal parameters determination, transmission of the tsunami message from the tsunami center to the civil protection, sending tsunami warning from the civil protection to the local population, and the local population evacuation, respectively. From numerical simulations as well as from the Tsunami

Travel Time Mapping Tool calculations (Figure 7.4) it has been found that $t_{tr} \leq 15$ min for tsunami sources situated at distances no more than about 100 km from the city of Rhodes which lies at the northernmost tip of the island. For earthquake magnitudes of about 7.0 or 7.2, which is the case of Rhodes large earthquakes, such as the 1957 one, the rupture length is about 80 km, thus justifying our assumption that near-field for Rhodes means tsunami travel distances not exceeding about 100 km. This is consistent with the definition proposed by Lauterjung et al. (2010) as mentioned previously.

According to the current status of the Hellenic National Tsunami Warning Center (HL-NTWC) at NOA and of the other national centers operating in the NEAM region, a strong earthquake is determined automatically by the seismograph system within about 3–5 min or manually by seismologists within about 10–20 min from the earthquake generation at best. Both modes of earthquake determinations are preliminary ones. Automatic solutions suffer from errors that usually are larger than the manual ones. Practically speaking the sum of t_{ED} and t_{TM} is no less than about 10 min. After the arrival of the tsunami message to the civil protection more time is needed to warn local people. Unfortunately, since the tsunami operational experience in the EM region is extremely limited so far, no data exist to calculate the time components t_{TW} and t_{EV}. Let us assume that $t_{TW} = 5$ min and $t_{EV} = 10$ min, although these estimates look like very optimistic. Then, we get that the time needed for evacuation exceeds by at least 10 min the first tsunami wave arrival time. This result underlines the urgent need to compress drastically the time needed for warning and real evacuation by developing a local tsunami early warning system.

The basic concept for the development of REWSET is that in order to gain in time we may lose in accuracy. Operationally this is translated to immediate civil protection mobilization as soon as a strong, potentially tsunamigenic earthquake has been instrumentally detected before its focal parameters are more accurately determined by the national seismograph center. The geophysical principle to initiate immediate response to an earthquake is the difference in arrival of the secondary, *S*, wave after the primary, *P*, wave. This difference increases with the increase of the epicentral distance. The SAD network is capable to detect both *P* and *S* waves. The recorded characteristics of the *P* wave permits drawing of conclusions as to the strength of the following *S* wave and, consequently, to the expected level of peak ground acceleration. If a preselected threshold of ground motion is exceeded, the SAD system may trigger alarms. The threshold level of ground motion is selected to correspond to an earthquake magnitude threshold.

The SAD system installed in Rhodes consist of *sectylifePatron*® devices produced by the *sectyelectronics GmbH* (Castrop-Rauxel, Germany), and tested in collaboration with GFZ (GeoforschungsZentrum, Potsdam, Germany), the coordinating institute of the EU-FP6 research project Seismic Early Warning for Europe (SAFER, 2006–2009). Such devices were also tested in NOA in the frame of SAFER under the author's supervision. SADs of this type have been installed for operational seismic early warning purposes in critical buildings (e.g., administrative centers, hotels, etc.) in several seismogenic countries around the world including Greece. To the author's knowledge REWSET is the first system to utilize SADs not only for seismic but also for tsunami early warning, thus conforming to recommendations internationally adopted for the multihazard orientation of tsunami early warning systems.

7.6.1.3 Operational Performance

Each SAD station installed in Rhodes Island consists of one master sensor and a slave sensor, which are placed at different installation points of the same building (Figure 7.6b). Seismic alert happens only when both the master and slave sensors in one SAD unit are triggered simultaneously, thus securing that the false alarm rate becomes minimal. Each one of the master and slave sensors is equipped with sirens and optical signal devices. An optimization study was performed as regards the triggering thresholds of SADs for strong earthquakes having their epicenters at a distance up to about 100 km from the city of Rhodes. Empirical relations between earthquake magnitude and peak ground acceleration were used to optimize triggering thresholds given that the devices installed permit to select from a scale of triggering thresholds. As soon as a SAD unit is activated by a strong earthquake, the device resets automatically in 30 s, thus becoming ready to receive new triggering signal. SAD units do not require special maintenance or needs. A usual minor problem is that they may shut down if the energy supply is interrupted. However, SAD units are autonomous from the energy supply point of view. In fact, normally they get energy from the central power supply system but in case of strong earthquake or other emergency, they are also supported by uninterruptible power supply devices.

Early detection of a strong earthquake occurring at distance up to 100 km from the city of Rhodes initiates emergency procedures in less than 25 s from the earthquake generation, depending on the epicentral distance. Then, civil protection and other local authorities are able to mobilize immediately and start actions aiming to start emergency procedures well before an automatic or manual preliminary determination of the earthquake parameters will be available by the national center some minutes after the earthquake generation.

Since the operation of SAD in Rhodes (July 2012) and the full operation of REWSET (December 2013), no strong earthquake occurred so far within the range of 100 km from Rhodes. On the other hand, no false alarm has been noted. From the point of view of testing the performance of the SAD network installed in Rhodes it is also of interest to examine the case of the strong earthquake that occurred to the east of Rhodes Island at 15:21:06 UTC of December 28, 2013 and measured magnitude 5.9 according to the earthquake solutions provided by the European–Mediterranean Seismological Center (EMSC). The epicenter location falls offshore Antalya at distance of about 208 km NW from Nicosia, Cyprus Island, and about 280 ESE from the city of Rhodes (Figure 7.9). The earthquake focal depth was ~65 km. From the distribution of the felt intensities (Figure 7.9), one may conclude that in Rhodes Island the ground motion shaking was very weak. In fact, none of the SAD stations were reported to be triggered by that earthquake. Even if the lowest triggering threshold had selected the SAD sensors would not have been triggered. For epicentral distance of about 280 km earthquake magnitude exceeding 6.5 is required for triggering one or more SAD units in Rhodes.

7.6.1.4 Future Plans

Future plans for the improvement of the REWSET may include the production of presimulated tsunami scenarios, based on the past earthquake and tsunami history and the active tectonics of the area, installation of additional SED units and tide gauges as well as tsunameters offshore Rhodes and surveillance cameras to control selected coastal zone segments immediately after a strong earthquake.

The REWSET system does not incorporate early detection of tsunamis caused by aseismic landslides, which are still out of control. Nevertheless, aseismic tsunamis are locally powerful and, therefore, quite hazardous. For example, the last case of March 26, 2002 in Rhodes was characteristic (Papadopoulos et al., 2007a). To contribute in this critical aspect an investigation of the prospects offered by the hydroacoustic technology for the early detection of such tsunamis deserves to be examined as a promising future tool.

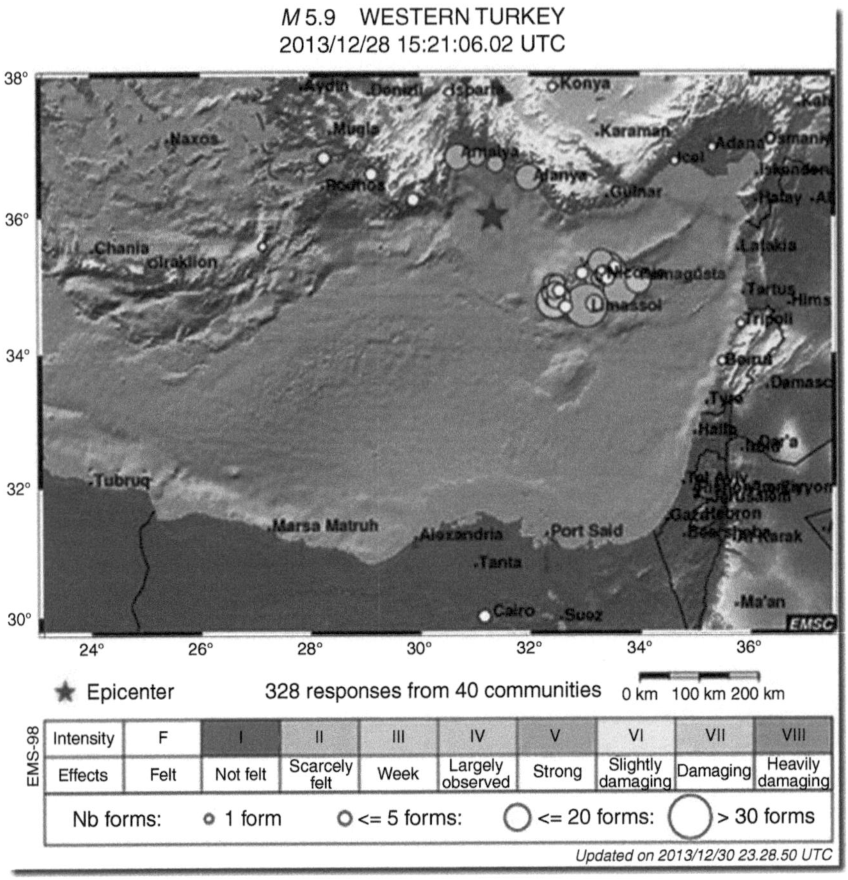

Figure 7.9 ***The epicenter (star) and intensity felt (circles) of the strong (M 5.9) earthquake of December 28, 2013 in the eastern Mediterranean (after EMSC).***

7.6.2 The Western Hellenic Arc Initiative

In Greece, the effort for the establishment of an experimental tsunami warning system in the western segment of the Hellenic Arc in the 1990s, already mentioned in section 7.2, was revitalized in the years 2006–2008 with the relevant national project ΕΝΙΣΧΥΣΗ (ENISXYSH-REINFORCEMENT) of NOA funded by the General Secretariat for Research and Technology, Greece. This project gave the opportunity for new or upgraded seismograph, tide gauge and GPS stations to be installed in western Crete. Pre-operational signal transmission to NOA of seismic and sea level signals was achieved via satellite Internet for the performance of monitoring and evaluation of the seismicity and of the sea level changes.

GPS data were transmitted to the Laboratory of Geodesy, Technical University of Crete, for long-term crustal deformation studies. The existence of this system practically contributed to accelerate the decision of the Greek Government taken in 2010 for the establishment by law of the Hellenic National Tsunami Warning Center at NOA, which became operational in the summer of 2012 not only at national level but also in the frame of NEAMTWS.

In the area of SW Peloponnese, Greece, situated at the westernmost sector of the Hellenic Arc, the area of Pylos town was selected as test site for the EU FP6 SEAHELLARC project (2007–2009) (Papoulia et al., 2014). The tsunami potential in the area is not negligible, as it is concluded from the existence of local and distant tsunami sources that threaten the area (Papadopoulos et al., 2014c; Yalciner et al., 2014). A pilot real-time on/offshore seismic and tsunami network, consisting of seismic stations and one tsunami pressure sensor, was installed in the Kyparissiakos Gulf, SW Peloponnese, transmitting data by satellite and GPRS (mobile telephony) connection, and exploiting the Internet possibilities (Makris et al., 2014). According to Makris et al. (2014), the marine real-time seismic and tsunami station, originally developed for a microseismicity study in central Greece, was further improved during the SEAHELLARC project.

7.6.3 Local Tsunami Early Warning System in Stromboli Island (Italy)

After the volcanic activity, that started on December 28, 2002, and the tsunamigenic landslide triggered downslope the cone along the scar known as "Sciara del Fuoco" on December 30, 2002 (Tinti et al., 2003; for more details and references see in Chapter 2), a multiparameter network was developed for the monitoring of the volcano by seismic, infrasonic and thermal signals (Ripepe and Marchetti, 2004). In addition, the ground deformation and the landslide mass are monitored by GPS and InSAR technologies.

The University of Florence, Department of Science, on behalf of the Italian Civil Protection Department, has positioned off the coast of Stromboli, a resinex elastic beacon equipped with an instrumental system with the aim to monitor the stability of the "Sciara del Fuoco." The beacon is anchored some 200 m off the coast at a depth of 43 m. The set of instruments includes a large band hydrostatic pressure sensor, a water temperature sensor, a hydroacoustic sensor positioned at a 14 m depth, two tiltmeters, and a GPS installed on the tower.

The beacon is able to provide immediately variables of hydrostatic pressure and monitoring of sea level fluctuations. The hydrophone is used to register possible hydroacoustic waves resulting from the underwater sliding of the volcano and which propagates in the water at speeds of 1500 m/s. The GPS receiver ensures the correct temporal synchronization of all data on board and monitors the geographical position.

The signals transmitted by the beacon also contain a series of diagnostic parameters, which enable the immediate and continuous monitoring of the on-board instrumentation and timely maintenance procedures. The entire monitoring system is powered by solar panels but has the capacity to operate autonomously for about 6 months without the aid of the panels.

From the Resinex beacon, the data are transmitted in "real time" via radio modem to the Civil Protection Advanced Operations Centre (COA) in Stromboli. The data are also sent via Internet to the civil Protection in Rome and to the University of Florence, Department of Science. For the alert of the local population in Stromboli, a system of sirens has been installed along the coastal area. This system is triggered when seismic and GPS sensors detect landslide occurrence and confirmation for tsunami generation comes from the wavemeter and hydroacoustic sensors. Activities for risk mitigation were complemented by a civil protection exercise that took place on April 19, 2005.

The previously mentioned system in Stromboli very likely was the first, even local, tsunami warning system that operated in the EM region, and this happened thanks to the landslide tsunami event of December 30, 2002 that "warned" of the urgent need to develop such a system. In fact, that event caused less destruction since it happened off-season. Had it struck on-season loss of life was likely unavoidable.

Improvements to the Stromboli early tsunami warning system have also been suggested. For example, from the performance of numerical simulations of potential landslide tsunami generation, Tinti et al. (2009) concluded that others than the "Sciara del Fuoco" tsunamigenic landslide sources exist and, therefore, the tsunami warning system in Stromboli should consider this possibility as well. On the other hand, Bellotti et al. (2009), taking the Stromboli 2002 case as an example, suggested that experimental tsunami simulation findings may offer opportunities for improving the warning system for landslide tsunamis in an island of quasicircular shape like Stromboli. Based on their results those authors discussed how and where it is convenient to place instruments for the tsunami waves detection. Similar suggestions were made by De Girolamo et al. (2014).

7.6.4 The French RATCOM Project

The RATCOM Project (Réseau d' Alerte aux risques Tsunamis et submersions COTières en Méditerranée) has been a multidisciplinary demonstration initiative funded by French ministerial and local authorities, aimed at developing a prototype, near-field tsunami warning system. This demonstrator is aimed at monitoring the tsunami hazard of the Ligurian coastal area, including the southeastern French to northwestern Italian and Corsican coastal areas. Tsunamis in this area could be generated by either local earthquakes or submarine landslides. This prototype system, dealing with the impact on the coasts of a tsunami generated in the near-field, has an architecture able to manage an end-to-end tsunami alert procedure, ranging from detection of the ongoing event to the population warning.

The virtual system is composed of two major functional components (Renou et al., 2011). *Upstream component*: the system relies on the acquisition of seismological and oceanographic data that are automatically processed and compared to precomputed scenarios in a data center in order to provide in a short time an assessment of the tsunami threat by mapping water heights and arrival times at pre-established coastal points. An effort has been made to minimize the rate of false alarm. Among the several tsunami scenarios, the collapse of the Nice airport embankment in 1979 that generated a tsunami (Chapter 2) has been a case of detailed study as far as the alert and the coastal impact are concerned. *Downstream component*: the rapid diffusion of the alert bulletin at a local scale is insured by a powerful and secured communication network further linked to larger alert systems including mass media for population warning: sirens (national alert network), mobile network using various technologies, including satellites, radio, TV, and electronic billboards.

7.6.5 Plans for Local Tsunami Early Warning in Western Norway

In western Norway, the tsunami risk is closely associated with landslides and rock falls in fjords. Information on historical events and relevant references can be found in Chapter 2. The most recent case was reported on January 29, 2014 when a coastal landslide occurred at Statland, Namdalseid county, mid-Norway, generating a local tsunami. Although the landslide and tsunami did not cause any human casualties, the induced tsunami gave rise to a considerable local runup height up to 10 m and local damage to Statland village (Glimsdal et al., 2014).

In the frame of the "Åknes/Tafjord Beredskap IKS Project", Harbitz et al. (2014) noted that an unstable rock volume of more than 50 million cubic meters has been detected in the Åkerneset rock slope in the narrow fjord, Storfjorden, Møre and Romsdal County, Western Norway. If large portions of the volume are released as a whole, the rockslide will generate a tsunami that may be devastating to several settlements and the numerous visiting tourists along the fjord. Harbitz et al. (2014) presented results from various numerical modeling and laboratory experiments and determined tsunami hazard zones based on results regarding the expected runup. During the risk assessment it was found that the rockslide tsunami hazard (probability of impact) is higher than accepted by the Norwegian Planning and Building Act. The Act is today altered to open for specified further development in the various hazard zones. The results of the tsunami analysis are applied in risk management in terms of hazard map production and land-use planning. Two failure scenarios for each of the two unstable rock slopes are designed for the hazard zoning. The larger and less probable scenarios (1 in 5000 years) are applied for evacuation zones and routes, while the smaller and more probable scenarios (larger than 1 in 1000 years) are applied for location and design of less critical facilities accepted in the inundation zone.

A monitoring system of the unstable rock volume is in place, consisting of various measurement and recording devices, such as rod extensometers, differential GPS, instrumented boreholes, two single lasers for distance measurements across the uppermost extension cracks, periodic laser scanning (ground-based LiDAR), in summertime a LISALab ground-based Interferometric Synthetic Aperture Radar (InSAR), satellite-based radar, a seismic network of geophones, an automated total station with 30 prisms, a weather station, and web cameras (ÅTB, 2010a; Blikra, 2012). Also, high-resolution topographic data were obtained from an airborne LiDAR campaign. In addition to, or as an alternative of, the monitoring systems in unstable rock slopes, Harbitz et al. (2014) supported that a system for rapid tsunami early warning by seabed pressure sensors should be elaborated. Such sensors will be able to detect all possible tsunamis in a fjord or a lake, not only those emerging from a release in already monitored rock slopes. They can also be applied to confirm or cancel warnings issued by other less reliable systems, and will provide valuable records in case of a real full-scale event. According to Harbitz et al. (2014), even though the warning time in fjords and lakes is short, ranging from a few to some tens of minutes, such a system will be significantly better than no warning at all.

7.6.6 The Adriatic Meteotsunami Network

Middle Adriatic is an area where the most destructive of the Adriatic meteotsunamis occur (Chapter 2). This network is located in the middle Adriatic, between Zadar and Dubrovnik, and being in a pilot phase, running continuously since March 2009 initially planned for research purposes. This network consists of three microbarograph meteorological stations, and is supplemented by four permanent tide gauge stations. Air pressure is measured each second; data are stored on an embedded computer and sent every hour to a central processing unit in Split where 1 min averages of data and 5 min rates of air pressure change are calculated. This is a simple, low cost meteotsunami research network created with the intention to upgrade it to an operational warning network (Šepić and Vilibić, 2011).

7.7 TSUNAMI EARLY WARNING AND RISK MITIGATION: SOME CRITICAL ISSUES

After more than 60 years of experience of tsunami early warning in the Pacific Ocean and in Japan, it comes out that the systems that have been operational so far were effective at an adequate level only as regards the early warning in the far-field domain. In fact, for distances of more than 2 or 3 h of tsunami travel time there is enough time for accurate forecasting of the wave arrival time and height in certain coastal forecast points. This is possible particularly if the wave has been firstly recorded by sea bottom sensors, as it happens in the last 15 years or so with the DART system in the Pacific Ocean and more recently in other oceans as well. Even in the near-field domain, the initial warning messages could be improved after such records become available. This is exactly the case of the large Tohoku tsunami of March 11, 2011 discussed in Chapter 1.

In the EM region, the main physical tsunami features, which are critical from the early warning point of view are the next (see more in Chapters 2–6): (a) the tsunami sources are situated as a rule very close to coastal zones, hence the hot issue is that tsunamis threaten coastal zones in the near-field domain; (b) most of the tsunamis are local but they are still powerful, and therefore, destructive waves; (c) large, basin-wide tsunamis, threatening coasts in both the near- and far-field domains, are infrequent but the risk is still high due to the high exposure and vulnerability of the coastal communities to the tsunami hazard; and (d) most local and basin-wide tsunamis are caused by earthquake activity but other generation mechanisms related to volcanic, landslide and rock fall processes should not be neglected.

After 2004 the building up of the NEAMTWS has been the main initiative for tsunami early warning in the European–Mediterranean level. As discussed earlier, this system became operational by the summer of 2012 after careful preparation and testing under the extremely helpful umbrella of IOC-UNESCO and the dedicated commitment of a long number of scientists and decision makers from more than 30 country members, all these people working on a volunteer basis.

The time interval of nearly 7 years between the establishment of the ICG/NEAMTWS (November 2005) and the initiation of the operational phase in summer of 2012 looks long. On November 5, 2008, 1 day before the annual ICG/NEAMTWS assembly in Athens, Greece, a voice of criticism from a tsunami scientist was saying: *Four years after a mega-tsunami killed more than 250,000 people in Asia, the Mediterranean remains the only world ocean or sea unprotected by any warning system.... Despite significant research funding – or perhaps because there is so much funding to be had – there is provocative inaction as Europe wastes time debating how many regional centers it needs, where to put them and how to name them. Old-fashioned scientific rivalries are major stumbling blocks to finding a quick and effective solution* (C. Synolakis, The Wall Street Journal, November 5, 2008).

However, the author of that statement did not take into account that in the EM region one has to bring together and coordinate more than 30 country members interested in tsunami early warning, each one being different from the others as regards the level of scientific and technological development, language, cultural background, and national administrative and civil protection regulations. This is also the case in other oceans. Therefore, the progress noted in the EM region for building up the NEAMTWS has been more or less parallel with the building up of the Indian Ocean Tsunami Warning System (IOTWS) and the Caribbean Tsunami Early Warning System (CARIBE EWS). For example, in the IOTWS, since 2012 the regional tsunami centers of Australia, India and Indonesia are the principal source of tsunami advisories for the Indian Ocean, while more than ten national centers operates and contributes in the overall system. This discussion is useful for someone, who is not involved in operational procedures, to understand that building up an operational system, like the NEAMTWS one, needs extremely careful preparation and testing from many points of view.

The architecture of the NEAMTWS is oriented to provide tsunami early watch services for tsunamis generated only by earthquakes. The operational experience existing three years after NEAMTWS entered its operational phase, indicates that after a strong earthquake which is potentially capable to produce tsunami the minimum time needed for the tsunami centers to send tsunami messages is about 15 min from the earthquake origin time. From this point of view it is quite questionable if NEAMTWS, in its present development, is efficient to provide early watch services in the near-field domain. This is exactly the case with the similar systems operating in the Pacific and Indian Oceans and in the Caribbean Sea. On the other hand, in the current practice the tsunami message is produced on the basis of the decision matrix (DM), an empirical tool (Tables 7.1 and 7.2), which is in immediate use after earthquake generation. However, the DM tool does not guarantee reliability and accuracy. Therefore, two relevant issues are how to improve (a) the early determination of the earthquake focal parameters and (b) the DM tool, in a way that may provide more reliable and accurate information about the tsunamigenic potential of the earthquake. An alternative to the use of the DM tool is the construction of databases of presimulated tsunami scenarios. As regards this point a critical issue is the selection of the best procedure that should be followed.

The detection of the tsunami itself by sea level recording systems yields verification that a tsunami has been really generated. The usefulness of the coastal tide gauges has been discussed in a previous section of this chapter. It was also noted that in the frame of NEAMTWS no sea bottom tsunami sensors have been operationally integrated so far. It is of particular value, therefore, to examine to what extent such sensors would be useful in the tsunami watch operations. Last but not least, is the issue of the communication infrastructures needed to facilitate tsunami early warning procedures.

The previously mentioned issues are among the most important concerning the improvement of the tsunami early warning operating in the EM region at regional and local scales. In the next lines these issues are discussed by taking into account the suggestions and improvements proposed by several tsunami scientists working in the EM region. New emerging methodologies and technologies are of particular values in this discussion.

7.7.1 Early Earthquake and Tsunami Determinations

One remarkable development in this field is the so-called *Early-Est (EE)* software, which is currently installed at the Centro Allerta Tsunami (CAT), the operational component of the Italian National Tsunami Warning Center (It-NTWC), in the seismic monitoring center of the Istituto Nazionale di Geofisica e Vulcanologia (INGV) in Rome (Italy). The concept, as explained by Bernardi et al. (2014), is that after a potentially tsunamigenic strong earthquake there is no time for a detailed analysis and accurate revision of the automatic earthquake solution, and only a quick validation/rejection of the results may be performed in most of the cases by a seismologist. Within this context, EE operates on continuous, real-time seismic waveform data to perform trace processing and picking, phase association, event detection, hypocenter location, and event characterization. This characterization includes m_b and M_{wp} magnitudes, and the determination of duration, T_0, of large earthquake magnitude, M_{wpd}, and assessment of the tsunamigenic potential based on some critical signal durations. Earlier introduction in the methodology can be found in a series of relevant papers including those of Lomax et al. (2007) and Lomax and Michelini (2011, 2012).

In order to test the performance of the fully automatic EE solutions for tsunami early warning, Bernardi et al. (2014) compared the hypocenters and magnitudes provided at global scale by different agencies for events of $M_w \geq 5.5$. Then, they compared the empirical uncertainties obtained in this way with EE solution and with the differences between the EE system and the reference catalogs. The analysis showed that EE is suitable for the purpose of the CAT since it generally provides fully automatic reliable locations and magnitudes within the uncertainties expected from statistical analysis of the manually revised reference catalogs. The time needed for obtaining such earthquake solutions are estimated to be around 10 min. Further improvements of the EE performance were suggested recently by Lomax et al. (2015).

Schindelé et al. (2015) explained that several methods were implemented at the CENALT (France) national tsunami center with the aim to compute the moment magnitude and the focal mechanism based on broad-band seismic records. In particular, they referred to the *W-phase* method developed by Kanamori and Rivera (2008), which for earthquake magnitudes in the range 6.5–9.0 provides accurate results in less than 40 min and that it is suitable for application in the NEAM region. This method, as well as the EE one, certainly may shorten the time needed for tsunami decision-making based on early earthquake parameter determination. However, it is again questionable if they are capable to support tsunami early warning in the very near-field domain where tsunami travel times are on the order of 10–15 min or even less.

Of relevance is also the Tsunami Early Detection Algorithm (TEDA), which has the goal to detect tsunami waves or high-amplitude long-period waves that can be dangerous to people, harbors, and properties (Bressan and Tinti, 2011). It is composed of two parallel algorithms: the TEDA "tsunami detection," which is focused on the identification of tsunami waves and is based on a signal slope discriminating approach and the TEDA "secure detection," which is based on incremental sea level estimation. Both algorithms work at station level, which has the advantage to use functions updated at every new data acquisition, without the need of waiting for a great load of data to make a decision on a possible alert (Bressan and Tinti, 2011).

Another important issue is the Faster Than Real Time (FTRT) tsunami simulation. The concept of FTRT has been approached by Macias et al. (2014) with the HySEA numerical model for the simulation of earthquake-generated tsunamis. The initial sea surface deformation is computed using Okada model. Wave propagation is computed using nonlinear shallow water equations in spherical coordinates, where coastal inundation and runup are suitably treated in the numerical algorithm. The generation, propagation and inundation phases are all integrated in a single code and computed coupled and synchronously when they occur at the same time. Inundation is modeled by allowing cells to dynamically change from dry to wet and reciprocally when water retreats from wetted areas.

Special effort has been made by Macias et al. (2014) in preserving the model well balanced (i.e., capturing small perturbations to the steady state of the ocean at rest). Such a graphics processing unit (GPU) model allows for a FTRT simulation for real large-scale problems. The large accelerations obtained make HySEA code suitable for its use in tsunami early warning systems. The Italian It-NTWC at INGV (Rome) has adopted HySEA GPU code. The model is verified by hind-casting the wave behavior in several benchmark problems. Numerical results for earthquake-generated tsunamis in the Mediterranean Sea is presented and computing time analyzed. This concept was further examined by Vida et al. (2015) with the aim to increase the accuracy of tsunami forecast by assimilating the largest possible amount of data in quasi real time, and performing simulations in a few minutes wall-clock time, possibly including the coastal inundation stage. According to Vida et al. (2015), this strategy of direct real time computation, that could seem unfeasible a decade ago, it is now foreseeable thanks to the astonishing recent increase in the computational power and bandwidth evolution of modern GPUs.

7.7.2 Is the Decision Matrix a Satisfactory Tool?

The decision matrix (DM) tool has been in use in the Pacific Tsunami Warning System (PTWS) since the 1960s. A similar tool was applied by JMA in Japan for several years. DM has been proved to be a useful tool in the absence of other, more reliable information about the generation or not of a tsunami immediately an earthquake occurrence. The verification of a tsunami occurrence comes from tsunami records, either in coastal tide gauges or in sea bottom sensors. Today the operational use of sea bottom tsunami sensors is a reality in the Pacific Ocean, including Japan, as well as in other oceans. However, the operational usefulness of such recording systems strongly depends on the near- or far-field character of the tsunami. The 2011 Tohoku tsunami in Japan was a near-field one for NE Japanese coasts but a far-field for remote places of the Pacific Ocean, for example, California. The problems faced with successful warning in the near-field domain, and the efforts undertaken to improve the warning systems in Japan, were discussed in Chapter 1. As regards the far-field domain, records by sea bottom sensors (e.g., DART system) provide adequate information for issuing successful tsunami forecasting hours ahead before the first arrival.

In the NEAM region, the urgent need concerns the very near-field tsunami warning, given that nearly all the known tsunami sources are situated very close to coasts. The distances typically range from 5 min to 30 min of tsunami travel time. This fact makes it unrealistic to wait for the tsunami verification from tide gauge records, since the coastal zone from which the first records are available are not warned. First records may be useful to warn more distant zones but again within very strict time constraints. On the other hand, as noted in a previous section, no sea bottom tsunami sensors are in place so far to support the NEAMTWS warning operations. Therefore, the DM, which is analyzed in a previous section is a useful tool justifying the policy of the ICG/NEAMTWS to adopt it as a basic operational tool.

A discussion that has opened within the ICG/NEAMTWS in the last years regards the efficiency of the DM tool adopted (Tables 7.1 and 7.2) as well as its possible improvements. For example, Tinti et al. (2012) showed that in Italy the actions prescribed by the DM are adequate only in 45–55% of the cases, overestimations are about 37% and underestimations are the rest. As a whole, the predictive ability of the DM is not satisfactory, which implies that recipients have the difficult task in managing bulletins carrying a great deal of uncertainty and, on the other hand, also suggests that strategies to improve the DM or to go beyond the DM need to be found (Tinti et al., 2012).

One of the characteristics of the DM, which is in use for the operational needs of the NEAMTWS is its gross regionalization, which accepts one DM for the North East Atlantic Ocean and another for the Mediterranean and connected seas. I support that further regionalization is possible and realistic. For example, in the Mediterranean Sea the definition of a tsunami being of possible regional impact (distance up to 400 km from the earthquake epicenter) may lead to an operational paradox. In fact, let's suppose the occurrence of a large ($M > 6.5$) earthquake in the Ligurian Sea leading to the issue of a watch-type tsunami message in a regional scale. This means that the warning of watch-type should apply to the North Adriatic Sea as well. However, it is hardly explainable how such a tsunami having its source in the Ligurian Sea could affect North Adriatic Sea by the severity predicted by the DM tool. Similarly, according to the DM tool a North Aegean Sea tsunami should affect also with the same severity the area of North Ionian Sea, for example, Corfu Island, which again is hardly explainable how it could happen. Therefore, a further regionalization of the existing DM tool would lead to more realistic and operationally applicable solutions.

Also, problems are created by considering as possible tsunami sources the earthquakes that have their epicenters at distances, d, up to 40 km inland. In high seismicity areas, for instance in Greece and western Turkey, such a consideration increases drastically the rate of tsunami messages that should be issued by the tsunami centers, which turns to a drastic increase of the tsunami false alarm rate. The decrease of the critical distance d is a solution but certainly this should be done carefully, for example, by taking d as a function of the earthquake magnitude.

7.7.3 Tsunami Scenario Database Building Strategies

In general, two different approaches can be used when choosing the strategy to building a tsunami scenario database (see short review in Armigliato et al., 2013). The first consists in discretizing a given "tsunamigenic domain" with sources distributed on a regular grid of cells, independently on the presence of well-defined tectonic lineaments with known or assumed tsunamigenic potential in each individual cell. Tsunami scenarios are then computed for each grid cell and for different earthquake magnitudes starting from "standard" tsunami initial conditions, like a cosinusoidal bump whose geometrical characteristics are somehow linked to the earthquake magnitude. This is, for example, the approach adopted by the Global Disasters Alerts and Coordination System (GDACS), developed by the JRC (Ulutas et al., 2012) and incorporated in the TAT which is implemented by some of the national centers supporting NEAMTWS.

The second approach applies when sound hypotheses on tectonic lineaments and/or specific active and potentially tsunamigenic faults can be made. Armigliato et al. (2013) call this a "source-based" approach: the fault areas are tessellated with elementary faults of suitable extension and with focal mechanism coinciding with that of the parent source area. This is the approach adopted, for instance, by the US NOAA Tsunami Research Center "short-term inundation forecast" (SIFT) operational tool: unit sources are defined along known and potential tsunamigenic tectonic zones in the Pacific and Atlantic oceans, and tsunami scenarios are computed for each of these predefined earthquake sources. Tsunami initial conditions are computed by taking in full account the source fault geometry and assumed characteristic focal mechanism, for example, by means of the classical Okada's formulas (Okada, 1992). In the frame of the NEARTOWARN project (2012–2013) supported by the EC DG-ECHO, the UNIBO (University of Bologna) tsunami team adopted the second approach (Armigliato et al., 2013) for the creation of what they called the matching scenario database (MSDB). The underlying assumption is that we have, and can improve in the future, a proper knowledge of the main tectonic structures acting close to the areas of interest. In view of these considerations, Armigliato et al. (2013) showed how a MSDB could be established in the NEAM region by taking the examples of the NEARTOWARN test areas of Rhodes and Cyprus islands.

7.7.4 Are Sea Bottom Tsunami Sensors Operationally Useful in NEAM Region?

As noted earlier, the NEAMTWS is still not supported by stations of sea bottom tsunami sensors. This is due to a number of reasons. The first is the cost for the installation of such stations. For a DART-type station the cost is rather high per complete station, although cheaper solutions could be found. Another problem is the maintenance, which adds significantly to the annual man effort and cost. Supposing that national centers and/or international consortia would conclude to appropriate solutions as regards financial and maintenance issues, then one should examine the operational usefulness of integrating sea floor tsunami sensors in the existing infrastructures that support the national tsunami center and the NEAMTWS.

As a first approach one may look in Figure 7.4a–c where tsunami travel time contours indicate that the operational centers have only very short-times to warn civil protection authorities and the general public. Let us take the example of Rhodes Island, the test site of NEARTOWARN project where a near-field tsunami warning system has been installed and operates fully since December 2013. Several important seismic tsunami sources exist around the island of Rhodes (Papadopoulos et al., 2007a). Such sources are situated at distances of no more than about 100 km from the city of Rhodes, situated at the northernmost tip of the island, that is within the tsunami travel time contour of about 15 min (Figure 7.4a). Suppose that a tsunami source and a sea bottom tsunami sensor are lying at distances of 100 and 50 km offshore Rhodes, respectively. Then, the warn time that one may save is on the order of 6–7 min. However, tsunami sources exist also at distances less than 100 km. Taking distance of 10 km offshore for the sensor, the warn time that can be saved is further reduced by 2-3 min. However, such reduced times are operationally still crucial in the very near-field domain. Therefore, the installation of sea bottom tsunami sensors deserves to be examined but a cost-benefit analysis may be needed per case.

Also, an optimum geographical distribution of both coastal tide gauges and sea bottom sensors would be of value to be examined. After a survey of European sea level infrastructure (tide gauges, telemetry methods, ancillary information) conducted at the end of 2008 on behalf of several projects and programs, approximately 478 strategic tide gauges were found to be operational at that time, of which about three-quarters have near-real time data telemetry of various kinds (Woodworth et al., 2009). Around half of the gauges take part in real-time international data exchange. Woodworth et al. (2009) supported that the NEAMTWS network can be considered to be in good shape in that most of its sites for which a gauge exists will be capable of meeting required standards in the near future. On the other hand, NEAMTWS (and the European and North African network in general) contains major gaps along the North African coastline and on European, Mediterranean and Black Sea coasts, which require new installations. Although new installations were realized after the findings of that survey, gaps still exist. In an optimum design study, Schindelé et al. (2008) found that in the case of Western Mediterranean, a network of around 17 coastal tide gauges and 13 tsunameters (sea bottom sensors) located at 50 km along the shore is required to detect and measure nearly all tsunamis generated on the northern coasts of Africa.

7.7.5 Support of TEWS by Cloud Computing

Tsunami early warning systems (TEWS) should be supported by instant tsunami generation and propagation calculations, by retrieving presimulation scenarios from relevant databases, as well as by communication technologies. Wächter et al. (2012) and Hammitzsch et al. (2015) supported that based on the experiences and the knowledge gained in research projects, such as the FP6 and FP7 EU DEWS and TRIDEC projects, new technologies are exploited to implement a cloud- and web-based platform – the TRIDEC Cloud – to open up new prospects for EWS. The platform developed in TRIDEC in its current version addresses tsunami early warning and mitigation. It merges several complementary external and in-house cloud-based services for instant tsunami propagation calculations and automated background computation with graphics processing units (GPU), for web-mapping of hazard specific geospatial data, and for serving relevant functionality to handle, share, and communicate threat specific information in a collaborative and distributed environment.

The TRIDEC Cloud can be accessed in two different modes, the monitoring mode and the exercise and training mode. The monitoring mode provides important functionality required to act in a real event. So far, the monitoring mode integrates historic and real-time sea level data and latest earthquake information. The integration of sources is supported by a simple and secure interface. The exercise and training mode enables training and exercises with virtual scenarios. This mode disconnects real world systems and connects with a virtual environment that receives virtual earthquake information and virtual sea level data replayed by a scenario player. Thus, operators and other stakeholders are able to train skills and prepare for real events and large exercises. GFZ, KOERI, and IPMA have used the opportunity provided by the NEAMTWS exercise NEAMWave14 to test the TRIDEC Cloud as a collaborative activity based on previous partnership and commitments at the European scale. The TRIDEC Cloud has not been involved officially in Part B of the NEAMWave14 scenarios. However, the scenarios have been used by GFZ, KOERI, and IPMA for testing in exercise runs on October 27–28, 2014. Additionally, the Greek NEAMWave14 scenario has been tested in an exercise run by GFZ only on October 29, 2014.

According to Hammitzsch et al. (2015), the exercise runs demonstrated that operators in warning centers and stakeholders of other involved parties just need a standard web browser to access a full-fledged TEWS. The integration of GPU accelerated tsunami simulation computations have been an integral part to foster early warning with on-demand tsunami predictions based on actual source parameters. Thus tsunami travel times, estimated times of arrival and estimated wave heights are available immediately for visualization and for further analysis and processing. The generation of warning messages is based on internationally agreed message structures and includes static and dynamic information based on earthquake information, instant computations of tsunami simulations, and actual measurements.

Generated messages are served for review, modification, and addressing in one simple form for dissemination via Cloud Messages, Shared Maps, email, FTP/GTS, SMS, and FAX. Cloud Messages and Shared Maps are complementary channels and integrate interactive event and simulation data. Thus recipients are enabled to interact dynamically with a map and diagrams beyond traditional text information.

7.8 OTHER TSUNAMI RISK MITIGATION ISSUES

Methods developed for the tsunami risk assessment in the EM region were discussed in Chapter 6. Also, the early warning systems developed so far, the technologies that support them and the large exercises conducted in the last years were analyzed previously in this chapter. In the next lines a few examples of other risk mitigation actions are reviewed.

The role of NEAMTIC has been catalytic in the coordination and production of material (books, brochures, etc.) that support tsunami risk mitigation actions in the EM region. Such actions include the organization of operational tsunami warning procedures and protocols, the performance of tsunami exercises and drills, education and awareness, and more. Such materials are freely accessible through the NEAMTIC web site (http://neamtic.ioc-unesco.org/).

After the tragic experience of the Indian Ocean 2004 devastating tsunami, other tsunami risk mitigation actions started to be undertaken at national level in several EM countries. For example, Omira et al. (2013) investigated the effectiveness of sea-defense structures in preventing/reducing the tsunami overtopping as well as evaluating the resulting tsunami impact at El Jadida, Morocco. Using the tsunami scenario technique and numerical simulations, those authors examined the behavior of incident tsunami waves when interacting with coastal infrastructures. Tsunami impact at the affected site was assessed through computing inundation and current velocity using a high-resolution digital terrain model that incorporates bathymetric, topographic, and coastal structures data. The predicted results show different levels of impact that different tsunami wave conditions could generate in the region. Omira et al. (2013) supported that existing coastal barriers around the El Jadida harbor succeeded in reflecting relatively small waves generated by some scenarios, but failed in preventing the overtopping caused by waves from others. Considering the scenario highly impacting the El Jadida coast, significant inundations were computed at the sandy beach and unprotected areas. Omira et al. (2013) concluded that the modeled dramatic tsunami impact in the region shows the need for additional tsunami standards not only for sea-defense structures but also for the coastal dwellings and houses to provide potential in-place evacuation.

The important issue of developing evacuation plans was examined by Sahal et al. (2013) in the French Mediterranean shore (French Riviera). They noted that evacuation is to be considered along the French Riviera, but no plan exists within communities. Those authors showed that various approaches can provide local stakeholders with evacuation capacities assessments to develop adapted evacuation plans through the case study of the Cannes–Antibes region. The complementarity between large- and small-scale approaches was demonstrated with the use of macrosimulators (graph based) and microsimulators (multiagent based) to select shelter points and choose evacuation routes for pedestrians located on the beach. The first one allows automatically selecting shelter points and measuring and mapping their accessibility. The second one shows potential congestion issues during pedestrian evacuations, and provides leads for the improvement of urban environment. Temporal accessibility to shelters was compared to potential local and distal tsunami travel times, and Sahal et al. (2013) concluded that the results show a 40 min deficit for an adequate crisis management in the first scenario and a 30 min surplus for the second one.

Very recently, in the frame of the EU FP7 ASTARTE tsunami research project standardized questionnaire surveys were performed in several test sites of the project. The aim of these surveys was to investigate the level of tsunami risk awareness by the general public. Some results started already to appear. For example, in the highly touristic test site of Heraklion, Crete Island, Greece, preliminary results showed that Greek and foreign visitors are better aware about the tsunami risk than local people (Papageorgiou et al., 2015). It would be of great interest to see and compare results from other test sites as well.

APPENDIX

Strong Tsunamis Historically Known in the Mediterranean and Connected Seas

Tsunami events are included here only if they have assigned intensity, *K*, of at least 6 in the 12-grade scale of Papadopoulos and Imamura (2001) by (see in Chapter 1) and are of reliability of at least 3 in a 4-grade scale (for explanation about the reliability scale see in Tinti and Maramai (1996) and Papadopoulos (2003a)). For data sources and literature see in Chapter 2. Symbol key: (−), BC date; *K*, tsunami intensity; *h*, run up height; *ML*, Murty and Loomis (1980) tsunami magnitude (see more in Chapter 1).

No.	Year	Month	Day	Region	Area/K	h (cm)	ML
1	−426	Summer		Maliac Bay	8		
2	−373	Winter		W. Corinth Gulf	Helike 9		
3	148			Rhodes Island	Rhodes 7		
4	365	07	21	Crete Island	Alexandria 10		
5	447	01	26	Marmara Sea	8		
6	544			Bulgarian Black Sea	8–9		
7	551	07	09	Lebanon	8		
8	556			Cos Island	Cos 8		
9	749	01	18	Levantine coast	7		
10	1169	02	04	Messina Straits	8		
11	1202	05	20	Syrian coast and Cyprus	7		
12	1303	08	08	Crete Island	Heraklion 10		
13	1343	10	18	Marmara Sea	8	200	
14	1365	01	02	Algiers	8		
15	1389	03	20	Chios Island	6		
16	1402	06		Corinth Gulf	8		
17	1481	05	03	Rhodes Island	Rhodes 7		
18	1598	05		Turkish Black Sea	8–9		
19	1609	04		Rhodes Island	Rhodes 8		
20	1627	07	30	Gargano	6		−1.4

(Continued)

Tsunamis in the European-Mediterranean Region. http://dx.doi.org/10.1016/B978-0-12-420224-5.00008-9

No.	Year	Month	Day	Region	Area/K	h (cm)	ML
21	1650	10	11	Thera Island	Patmos 10	2000	+3.0
22	1693	01	11	Eastern Sicily	7		+2.3
23	1741	01	31	Rhodes Island	Rhodes 8		
24	1748	05	25	W. Corinth Gulf	Aeghion 9	1000	
25	1755	11	01	SW Iberia	Lisbon 10	1500	+3.8
26	1759	11	25	Levantine Sea	Akko 8		
27	1766	05	22	Marmara Sea	7		
28	1773	05	06	Tangiers	7	900	
29	1783	02	06	Calabria	9	900	−1.8
30	1817	08	23	W. Corinth Gulf	Aeghion 9	500	
31	1823	03	05	North Sicily	8		
32	1856	11	13	Chios Island	8		
33	1866	01	02	Albania	7		
34	1866	02	02	Kythira Island	Avlemonas 6	800	
35	1866	03	06	Albania	7		
36	1867	09	09	SE Peloponnese	Gythion 7		
37	1908	12	28	Messina Straits	10	1300	−0.4
38	1944	08	20	Stromboli Island	7		
39	1948	02	09	Karpathos Island	7		
40	1956	07	09	Cyclades	Astypalaea 9	1500	+3.0
41	1963	02	07	W. Corinth Gulf	7	500	−11.0
42	1979	04	15	Montenegro	8		
43	1999	08	17	Marmara Sea	6	250	
44	2002	12	30	Stromboli Island	Ficogrande 7	900	

REFERENCES

Abe, K., 1979. Size of great earthquakes of 1837–1974 inferred from tsunami data. J. Geophys. Res. 84, 1561–1568.

Abe, K., 1985. Quantification of major earthquake tsunamis of the Japan Sea. Phys. Earth Planet. Inter. 38, 214–223.

Abe, K., 1989. Quantification of tsunamigenic earthquakes by the Mt scale. Tectonophysics 166, 27–34.

Aki, K., 1966. Generation and propagation of G waves from the Niigata earthquake of June 16, 1964, Part 2. Bull. Earthq. Res. Inst. Tokyo Univ. 44, 73–88.

Alasset, P.J., Hébert, H., Maouche, S., Calbini, V., Meghraoui, M., 2006. The Tsunami induced by the 2003 Zemmouri earthquake (M_W = 6.9, Algeria): modelling and results. Geophys. J. Int. 166, 213–226.

Alpar, B., Ünlü, S., Altinok, Y., Özer, N., Aksu, A., 2012. New approaches in assessment of tsunami deposits in Dalaman (SW Turkey). Nat. Hazards 60, 27–41.

Altinok, Y., Alpar, B., Ersoy, Ş., Yalçiner, A.C., 1999. Tsunami generation of the Kocaeli Earthquake (August 17th 1999) in the Izmit Bay; coastal observations, bathymetry and seismic data. Turk. J. Mar. Sci. 5 (3), 131–148.

Altinok, Y., Alpar, B., Özer, N., Aykurt, H., 2011. Revision of the tsunami catalogue affecting Turkish coasts and surrounding regions. Nat. Hazards Earth Syst. Sci. 11, 273–291.

Altinok, Y., Alpar, B., Özer, N., Yaltirak, C., 2003. Şarköy-Mürefte 1912 earthquake's tsunami, extension of the associated faulting in the Marmara sea, Turkey. J. Seismol. 7, 329–346.

Altinok, Y., Ersoy, Ş., 2000. Tsunamis observed on and near the Turkish coast. Nat. Hazards 21, 185–205.

Ambraseys, N.N., 1960. The seismic sea-wave of July 1956 in the Greek Archipelago. J. Geophys. Res. 65 (4), 1257–1265.

Ambraseys, N.N., 1962. Data for the investigation of the seismic sea-waves in the Eastern Mediterranean. Bull. Seismol. Soc. Am. 52, 895–913.

Ambraseys, N.N., 1965. The seismic history of Cyprus. Revue pour l'etude des calamites 40, 25–48.

Ambraseys, N.N., 2009. Earthquakes in the Mediterranean and Middle East, A Multidisciplinary Study of Seismicity up to 1900. Cambridge University Press, Cambridge, UK, 947 pp.

Ambraseys, N.N., Melville, C.P., 1988. An analysis of the eastern Mediterranean earthquake of 20 May 1202. In: Lee, W.H.K., Meyers, H., Shimazaki, K. (Eds.), Historical Seismograms and Earthquakes of the World. Academic Press, San Diego, USA, pp. 181–200.

Amiran, D.H.K., Arieh, E., Turcotte, T., 1994. Earthquakes in Israel and adjacent areas: macroseismic observations since 100 B.C.E. Israel Explor. J. 44, 260–306.

Ammon, Ch.J., Lay, Th., Kanamori, H., Cleveland, M., 2011. A rupture model of the 2011 off the Pacific coast of Tohoku earthquake. Earth Planets Space 63, 693–696.

Amoruso, A., Crescentini, L., Neri, G., Orecchio, B., Scarpa, R., 2006. Spatial relation between the 1908 Messina Straits earthquake slip and recent earthquake distribution. Geophys. Res. Lett. 33, L17309.

Amoruso, A., Crescentini, L., Scarpa, R., 2002. Source parameters of the 1908 Messina Straits, Italy, earthquake from geodetic and seismic data. J. Geophys. Res. 107, B4, 2080, 10.1029/2001JB000434.

Amoruso, A., Crescentini, L., Scarpa, R., 2010. Comment on "The 28 December 1908 Messina Straits Earthquake (Mw 7.1): a great earthquake throughout a Century of Seismology" by Pino, N.A., Piatanesi, A., Valensise, G., Boschi, E. Seism. Res. Lett., 81, 225–228.

Tsunamis in the European-Mediterranean Region. http://dx.doi.org/10.1016/B978-0-12-420224-5.00016-8

Ando, M., Ishida, M., Hayashi, Y., Mizuki, Ch., 2011. Interviews with survivors of Tohoku earthquake provide insights into fatality rate. EOS Trans. AGU 92 (46), 411–412.

Andrade, C., Borges, P., Freitas, M.C., 2006. Historical tsunami in the Azores archipelago (Portugal). J. Volcanol. Geoth. Res. 156, 172–185.

Annunziato, A., 2007a. The tsunami assessment modelling system by the Joint Research Centre. Sci. Tsunami Hazards 26 (2), 70–92.

Annunziato, A., 2007b. The JRC tsunami assessment modelling system. JRC Science and Technical Reports, European Commission, 56 pp.

Annunziato, A., Carrilho, F., Matias, L., Baptista, M.A., Omira, R., 2009. Progresses in the establishment of the Portuguese tsunami warning system, EMSC Newsletter, April 2009, pp. 10–12.

Antonopoulos, J., 1973. Tsunamis of the Eastern Mediterranean From Antiquity to Present Time, Athens. (in Greek).

Antonopoulos, J., 1979. Catalogue of tsunamis in the Eastern Mediterranean from antiquity to present times. Ann. Geof. 32, 113–130.

Antonopoulos, J., 1980. Data for investigation on seismic sea-waves events in the eastern Mediterranean from the Birth of Christ to 1980 AD (six Parts). Ann. Geofis. 33, 141–248.

Antonopoulos, J., 1992a. The great Minoan eruption of Thera volcano and the ensuing tsunami in the Greek Archipelago. Nat. Hazards 5, 153–168.

Antonopoulos, J., 1992b. The tsunami of 426 B.C. in the Maliakos Gulf, Eastern Greece. Nat. Hazards 5, 83–93.

AOA, 1912. Annales des l' Observatoires d' Athènes.

Argnani, A.F., Chiocci, L., Tinti, S., Bosman, A., Lodi, M.V., Pagnoni, G., Zaniboni, F., 2009. Comment on "On the cause of the 1908 Messina tsunami, southern Italy" by Andrea Billi et al. Geophys. Res. Lett. 36, L13307.

Argus, D.F., Gordon, R.G., De Mets, C., Stein, S., 1989. Closure of the Africa-Eurasia-North America plate motion circuit and tectonics of the Gloria fault. J. Geophys. Res. 94, 5585–5602.

Armigliato, A., Pagnoni, G., Paparo, M.A., Tinti, S., Zaniboni, F., 2013. Numerical demonstrator for pre-simulated tsunami scenarios. Report on Deliverable DC1: NERA-TOWARN DG-ECHO Project, October 2013, 44 pp.

Armijo, R., Flerit, F., King, G., Meyer, B., 2004. Linear elastic fracture mechanics explains the past and present evolution of the Aegean. Earth Planet. Sci. Lett. 1–2 (217), 85–95.

Assier-Rzadkiewicz, S., Heinrich, P., Sabatier, P.C., Savoye, B., Bourillet, J.F., 2000. Numerical modelling of a landslide-generated Tsunami: the 1979 Nice Event. Pure Appl. Geophys. 157, 1707–1727.

ÅTB, 2010a. Åknes: state of instrumentation and data analysis. Åknes report 02.2010. http://aknes.no/sites/default/files/Aaknesrapport02_2010.pdf.

ÅTB, 2010b. Hegguraksla in Tafjorden: monitoring and data analysis. Åknes report 04.2010. http://aknes.no/sites/default/files/Aaknesrapport04_2010.pdf.

Atillah, A., El Hadani, D., Moudni, H., Lesne, O., Renou, C., Mangin, A., Rouffi, 2011. Tsunami vulnerability and damage assessment in the coastal area of Rabat and Sale. Morocco. Nat. Hazards Earth Syst. Sci. 11, 3397–3414.

Baptista, M.A., Miranda, M., 2009. Revision of the Portuguese catalog of tsunamis. Nat. Hazards Earth Syst. Sci. 9, 25–42.

Baptista, M.A., Miranda, M., Victor, L.M., 1992. Maximum entropy analysis of Portuguese tsunami data: the tsunamis of 28.02.1969 and 26.05.1975. Sci. Tsunami Hazards 10, 9–20.

Baptista, M.A., et al., 1998. The 1755 Lisbon Tsunami; evaluation of the tsunami parameters. J. Geodyn. 25 (2), 143–157.

Baptista, M.A., et al., 2003. New study of the 1755 earthquake source based on multichannel seismic survey data and tsunami modeling. Nat. Hazards Earth Syst. Sci. 3, 333–340.

Baptista, M.A., Miranda, J.M., Omira, R., Antunes, C., 2011. Potential inundation of Lisbon downtown by a 1755-like tsunami. Nat. Hazards Earth Syst. Sci. 11, 3319–3326.

Baratta, M., 1910. La catastrofe sismica calabro-messinese del 28 dicembre 1908. Relazione alla Società Geografica Italiana. 2, 84 (in Italian).
Baraza, J., et al., 1990. Geotechnical characteristics and slope stability on the Ebro margin, Western Mediterranean. Mar. Geol. 95, 379–393.
Bartolome, R., Gràcia, E., Stich, D., Martinez-Loriente, S., Klaeschen, D., Mancilla, F.L., Lo Iacono, C., Dañobeitia, J.J., Zitellini, N., 2012. Evidence for active strike-slip faulting along the Eurasia-Africa convergence zone: implications for seismic hazard in the SW Iberian Margin. Geology 40, 495–498.
Basili, R., Tiberti, M.M., Kastelic, V., Romano, F., Piatanesi, A., Selva, J., Lorito, S., 2013. Integrating geologic fault data into tsunami hazard studies. Nat. Hazards Earth Syst. Sci. 13, 1025–1050.
Beisel, S., Chubarov, L., Didenkulova, I., Kit, E., Levin, A., Pelinovsky, E., Shokin, Y., Sladkevich, M., 2009. The 1956 Greek tsunami recorded at Yafo, Israel, and its numerical modeling. J. Geophys. Res. 114, C09002.
Bell, K.L.C., et al., 2012. Submarine evidence of a debris avalanche deposit on the eastern slope of Santorini volcano, Greece. Tectonophysics 597/598, 147–160.
Bellotti, G., Di Risio, M., De Girolamo, P., 2009. Feasibility of tsunami early warning systems for small volcanic islands. Nat. Hazards Earth Syst. Sci. 9, 1911–1919.
Benchekroun, S., Omira, R., Baptista, M.A., El Mouraouah, A., Iben Brahim, A., Toto, E.A., 2013. Tsunami impact and vulnerability in the harbor area of Tangier, Morocco, Geomatics. Nat. Hazards Riskdoi: 10.1080/19475705.2013.858373.
Bernard, E.N., Robinson, A.R. (Eds.), 2009. In: Tsunamis. The Sea: Ideas and Observations on Progress in the Study of the Seas, vol. 15. Harvard University Press, Cambridge, MA, 450 pp.
Bernardi, F., Lauciani, V., Lomax, A., Lorito, S., Michelini, A., Piatanesi, A., 2014. Performance analysis of the Early-Est software within the tsunami early warning system installed at the INGV. Geophys. Res. Abstracts 16, EGU2014-14884-2, 2014, EGU General Assembly 2014.
Bertolaso, G., Boschi, E., Guidoboni, E., Valensise, G. (Eds.), 2008. Il Terremoto e il Maremoto del 28 dicembre 1908 analisi sismologica, impatto, prospettive. INGV-DPC, Roma-Bologna, 813 pp.
Bilek, S., Lay, T., 1999. Rigidity variations with depth along interpolate megathrust faults in subduction zones. Nature 400, 443–446.
Billi, A., Funiciello, L., Minelli, L., Faccenna, C., Neri, G., Orecchio, B., Presti, D., 2008. On the cause of the 1908 Messina tsunami, southern Italy. Geophys. Res. Lett. 35, 6.
Billi, A., Minelli, L., Orecchio, B., Presti, D., 2010. Constraints to the cause of three historical tsunamis (1908, 1783, and 1693) in the Messina Straits Region, Sicily, Southern Italy. Seismological Res. Lett. 81, 907–915.
Birkmann, J., Teichman, K.v., Welle, T., González, M., Olabarrieta, M., 2010. The unperceived risk to Europe's coasts: tsunamis and the vulnerability of Cadiz, Spain. Nat. Hazards Earth Syst. Sci. 10, 2659–2675.
Blanc, P.L., 2009. Earthquakes and tsunami in November 1755 in Morocco: a different reading of contemporaneous documentary sources. Nat. Hazards Earth Syst. Sci. 9, 725–738.
Blikra, L.H., 2012. The Åknes rockslide, Norway. In: Clague, J.J., Stead, D. (Eds.), Landslides: Types, Mechanisms and Modeling. Cambridge University Press, Cambridge, pp. 323–335.
Bondevik, S., Mangerud, J., Dawson, S., Dawson, A., Lohne, Ø., 2003. Record-breaking height for 8000-year-old tsunami in the North Atlantic. EOS, Transactions. AGU 84 (31), 289–293.
Bondevik, S., Svendsen, J.I., Johnsen, G., Mangerud, J., Kaland, P.E., 1997. The Storegga tsunami along the Norwegian coast, its age and runup. Boreas 26, 29–53.
Borges, P., Andrade, C., Freitas, M.C., 2008. Tsunamis históricos na ilha do Faial (Açores-Portugal). In: Oliveira, C.S., Costa, A., Nunes, J.C. (Eds.), Sismo 1998-Açores. Uma Década Depois. SPRHI, SA. Horta, pp. 707–716, (in Portugese with English abstr.).

Borrero, J.C., Synolakis, C.E., Fritz, H., 2006. Northern Sumatra field survey after the December 2004 great Sumatra earthquake and Indian Ocean tsunami. In: Iwan, W.D. (Ed.), The Great Sumatra Earthquakes and Indian Ocean Tsunamis of 26 December 2004 and 28 March 2005 Reconnaissance Report. Earthquake Spectra, Special Issue III, vol. 22. Publication Number 2006-05, pp. S93-S104.

Bourgeois, J., MacInnes, B., 2010. Tsunami boulder transport and other dramatic effects of the 15 November 2006 central Kuril Islands tsunami on the island of Matua. Zeitschrift für Geomorphologie 54 (3), 175–195.

Bozzano, F., Lenti, L., Martino, S., Montagna, A., Paciello, A., 2011. Earthquake triggering of landslides in highly jointed rock masses: reconstruction of the 1783 Scilla rock avalanche (Italy). Geomorphology 129 (3–4), 294–308.

Bressan, L., Tinti, S., 2011. Structure and performance of a real-time algorithm to detect tsunami or tsunami-like alert conditions based on sea-level records analysis. Nat. Hazards Earth Syst. Sci. 11, 1499–1521.

Bruins, H.J., McGillivray, J.A., Synolakis, C.E., Benjamini, Ch., Keller, J., Kisch, H.J., Klügel, A., Van der Plicht, J., 2008. Geoarchaeological tsunami deposits at Palaikastro (Crete) and the Late Minoan IA eruption of Santorini. J. Archaeol. Sci. 35, 191–212.

Bryant, E., 2008. Tsunami-The Underrated Hazard, second ed. Springer-Praxis, UK, p. 330.

Buforn, E., Sanz de Galdeano, C., Udías, A., 1995. Seismotectonics of the Ibero-Maghrebian region. Tectonophysics 248, 247–261.

Buforn, E., et al., 2004. Seismic sources on the Iberia-African Plate boundary and their Tectonic Implications. Pure Appl. Geophys. 161, 623–646.

Camerlenghi, A., Acosta, J., Canals, M., Wardell, N., 2009. Morphogenesis of the SW Balearic continental slope and adjacent abyssal plain, Western Mediterranean Sea. Internat. J. Earth Sci. 98, 735–750.

Camerlenghi, A., Urgeles, R., Fantoni, L., 2010. A database on submarine landslides of the Mediterranean Sea. In: Mosher, D.C., Moscardelli, L., Shipp, R.C., Chaytor, J.D., Baxter, C.D.P., Lee, H.J., Urgeles, R. (Eds.), Submarine Mass Movements and Their Consequences. Advances in Natural & Technological Hazards Research, vol. 28. Spinger, Dordrecht, The Netherlands, pp. 491–501.

Camilleri, D.H., 2006. Tsunami construction risks in the Mediterranean – outlining Malta's scenario. Disaster Prev. Manag. 15, 146–162.

Campos, M.L., 1991. Tsunami hazard on the Spanish coasts of the Iberian Peninsula. Sci. Tsunami Hazards 9, 83–90.

Capuano, P., De Natale, G., Gasparini, P., Pingue, F., Scarpa, R., 1988. A model for the 1908 Messina Straits (Italy) earthquake by inversion of levelling data. Bull. Seismol. Soc. Am. 78, 1930–1947.

Cattaneo, A., et al., 2010. Submarine landslides along the Algerian margin: a review of their occurrence and potential link with tectonic structures. In: Mosher, D.C., Moscardelli, L., Shipp, R.C., Chaytor, J.D., Baxter, C.D.P., Lee, H.J., Urgeles, R. (Eds.), Submarine Mass Movements and Their Consequences. Advances in Natural and Technological Hazards Research, vol. 28, Springer, Dordrecht, pp. 515–525.

Cherubini, P., Humbel, T., Beeckman, H., et al., 2014. The olive-branch dating of the Santorini eruption. Antiquity 88, 267–291.

Chiocci, F., Romagnoli, C., Bosman, A., 2008a. Morphologic resilience and depositional processes due to the rapid evolution of the submerged Sciara del Fuoco (Stromboli Island) after the December 2002 submarine slide and tsunami. Geomorphology 100 (3–4), 356–365.

Chiocci, F.L., Romagnoli, C., Tommasi, P., Bosman, A., 2008b. The Stromboli 2002 tsunamigenic submarine slide: characteristics and possible failure mechanisms. J. Geophys. Res. 113, B10102.

Choi, B.H., Pelinovsky, E., Kim, K.O., Lee, J.S., 2003. Simulation of the trans-oceanic tsunami propagation due to the 1883 Krakatau volcanic eruption. Nat. Hazards Earth Syst. Sci. 3 (5), 321–332.

Choi, E., DesRoches, R., Nielson, B., 2004. Seismic fragility of typical bridges in moderate seismic zones. Eng. Struct. 26, 187–199.
Christofidis, M.K., 1969. Oi seismoi kai ai seismikai doniseis en Cypro apo tis arxaiotitos mexri simeron. Cypriakos Logos, Cyprus, vol. 6, pp. 323–325.
CIESM, 2011. Marine geo-hazards in the Mediterranean, no. 42 in CIESM Workshop Monographs. In: Briand, F. (Ed.), 192 pages, Monaco.
Cita, M.B., Aloisi, G., 2000. Deep-sea tsunami deposits triggered by the explosion of Santorini (3500 years BP), Eastern Mediterranean. Sediment. Geol. 135, 181–302.
Cita, M.B., Camerlenghi, A., Kastens, K.A., McCoy, F.W., 1984. New findings of bronze age homogenites in the Ionian sea: geodynamic implications for the Mediterranean. Mar. Geol. 55, 47–62.
Cita, M.B., Camerlenghi, A., Rimoldi, B., 1996. Deep-sea tsunami deposits in the eastern Mediterranean: new evidence and depositional models. Sediment. Geol. 104, 155–173.
Coburn, A., Spence, R., 1992. Earthquake Protection. John Wiley & Sons, Chichester, 355 pp.
Commission Staff Working Paper, 2010. Risk assessment and mapping guidelines for disaster management. Euroepan Commission, Brussels, SEC(2010), 1626 final, 42 pp.
Cruz, A.M., Krausmann, E., Franchello, G., 2011. Analysis of tsunami impact scenarios at an oil refinery. Nat. Hazards 58, 141–162.
Daëron, M., Klinger, Y., Tapponnier, P., Elias, A., Jacques, E., Sursock, A., 2005. Sources of the large A. D. 1202 and 1759 Near East earthquakes. Geology 33, 529–532.
Dale, K., Flay, S., 2006. Structural vulnerability estimation for tsunami loads, Earthquake Engineering in Australia, Canberra, November 24–26, 2006, pp. 179–183.
Dall' Osso, F., Gonella, M., Gabbianelli, G., Withycombe, G., Dominey-Howes, D., 2009. A revised (PTVA) model for assessing the vulnerability of buildings to tsunami damage. Nat. Hazards Earth Syst. Sci. 9, 1557–1565.
Dall' Osso, F., Maramai, A., Graziani, L., Brizuela, B., Cavalletti, A., Gonella, M., Tinti, S., 2010. Applying and validating the PTVA-3 Model at the Aeolian Islands, Italy: assessment of the vulnerability of buildings to tsunamis. Nat. Hazards Earth Syst. Sci. 10, 1547–1562.
Dan, et al., 2007. The 1979 Nice harbour catastrophe revisited: trigger mechanism inferred from geotechnical measurements and numerical modeling. Mar. Geol. 245, 40–64.
Darawcheh, R., Sbeinati, M.R., Margottini, C., Paolini, S., 2000. The 9 July 551AD Beirut earthquake, Eastern Mediterranean region. J. Earthq. Eng. 4, 403–414.
Dawson, A.G., Hindson, R., Andrade, C., Freitas, C., Parish, R., Bateman, M., 1995. Tsunami sedimentation associated with the Lisbon earthquake of 1 November AD 1755: Boca do Rio, Algarve, Portugal. The Holocene 52, 209–215.
Dawson, A.G., Lockett, P., Shi, S., 2004. Tsunami hazards in Europe. Environ. Internat. 30, 577–585.
Dawson, A.G., Long, D., Smith, D.E., 1988. The Storegga slide: evidence from eastern Scotland for a possible tsunami. Mar. Geol. 82, 271–276.
De Girolamo, P., Di Risio, M., Romano, A., Molfetta, M.G., 2014. Landslide tsunami: physical modeling for the implementation of tsunami early warning systems in the Mediterranean Sea. Procedia Engineering, vol. 70, pp. 429–438.
De Martini, P.M., Barbano, M.S., Smedile, A., Gerardi, F., Pantosti, D., Del Carlo, P., Pirrotta, C., 2010. A unique 4000 year long geological record of multiple tsunami inundations in the Augusta Bay (eastern Sicily Italy). Mar. Geol. 276, 42–57.
De Mets, C., Gordon, R.G., Argus, D.F., 2010. Geologically current plate motions. Geophys. J. Int. 181, 1–80.
De Natale, G., Pingue, F., 1991. A variable slip fault model for the 1908 Messina Straits (Italy) earthquake, by inversion of levelling data. Geophys. J. Int. 104, 73–84.
Denton, F., Wilbanks, T.J., Abeysinghe, A.C., Burton, I., Gao, Q., Lemos, M.C., Masui, T., O'Brien, K.L., Warner, K., 2014. Climate-resilient pathways: adaptation, mitigation, and sustainable development. In: Field, C.B., Barros, V.R., Dokken, D.J., Mach, K.J., Mastrandrea, M.D., Bilir, T.E., Chatterjee, M., Ebi, K.L., Estrada, Y.O., Genova, R.C.,

Girma, B., Kissel, E.S., Levy, A.N., MacCracken, S., Mastrandrea, P.R., White, L.L. (Eds.), Climate Change 2014 Impacts, Adaptation, and Vulnerability. Part A: Global and Sectoral Aspects. Contribution of Working Group II to the Fifth Assessment Report of the Intergovernmental Panel on Climate Change. Cambridge University Press, Cambridge, UK and New York, NY, USA, pp. 1101–1131.

Diakogianni, G., Papadopoulos, G., Fokaefs, A., Papageorgiou, A., Triantafyllou, I., 2015. Space-time distribution of tsunami impact in the European-Meditterranean region as results from a new Tsunami Catalogue. Geophys. Res. Abstr. 17, EGU2015-8439, EGU General Assembly 2015.

Dias, W.P.S., Yapa, H.D., Peiris, L.M.N., 2009. Tsunami vulnerability functions from field surveys and Monte Carlo simulation. Civ. Eng. Environ. Syst. 26, 181–194.

Dominey-Howes, D., Dawson, A., Smith, D., 1998. Late Holocene coastal tectonics at Falasarna, W. Crete: a sedimentary study. In: Vita-Finzi, C. (Ed.), Coastal Tectonics, Geological Society London Special Publications, vol. 146, pp. 343–352.

Dominey-Howes, D., Papadopoulos, G.A., Dawson, A.G., 2000. Geological and historical investigation of the 1650 Mt. Columbo (Thera Island) eruption and tsunami, Aegean Sea, Greece. Nat. Hazards 21, 83–96.

Dominey-Howes, D., Papathoma, M., 2007. Validating a Tsunami Vulnerability Assessment Model (the PTVA Model) using field data from the 2004 Indian Ocean tsunami. Nat. Hazards 40, 113–136.

Dominey-Howes, D.P., Verner, J., Papathoma-Köhle, M., 2010. Estimating probable maximum loss from a Cascadia tsunami. Nat. Hazards 53, 43–61.

Dominey-Howes, D.T.M., 1996. Sedimentary deposits associated with the July 9th 1956 Aegean Sea tsunami. Phys. Chem. Earth 21, 51–55.

Dominey-Howes, D.T.M., 1998. Assessment of tsunami magnitude and implications for urban hazard planning in Greece. Disaster Prev. Manag. 7 (3), 176–182.

Dominey-Howes, D.T.M., 2002. Documentary and geological records of tsunamis in the Aegean Sea region of Greece and their potential application to risk assessment and disaster management. Nat. Hazards 25 (3), 195–224.

Dotsenko, S.F., Miklashevskaya, N.A., 2008. Thermohydrodynamics of the ocean generation of seiches by moving baric fronts in bounded basins. Phys. Oceanogr. 18 (2), 63–77.

Douglas, J., 2007. Physical vulnerability modelling in natural hazard risk assessment. Nat. Hazards Earth Syst. Sci. 7, 283–288.

Drab, L., Hubert Ferrari, A., Schmidt, S., Martinez, P., 2012. The earthquake sedimentary record in the western part of the Sea of Marmara, Turkey. Nat. Hazards Earth Syst. Sci. 12, 1235–1254.

Droz, L., dos Reis, A.T., Rabineau, M., 2006. Quaternary turbidite systems on the northern margins of the Balearic Basin (Western Mediterranean): a synthesis. Geo-Mar. Lett. 26, 347–359.

Ducassou, E., Migeon, S., Capotondi, L., Mascle, J., 2013. Run-out distance and erosion debris-flows in the Nile deep-sea fan system: evidence from lithofacies and micropaleontological analyses. Mar. Pet. Geol. 39, 102–123.

Ducassou, E., Migeon, S., Mulder, T., Murat, A., Capotondi, L., Bernasconi, S.M., Mascle, J., 2009. Evolution of the Nile deep-sea turbidite system during the Late Quaternary: influence of climate change on fan sedimentation. Sedimentology 56, 2061–2090.

Duputel, Z., Rivera, L., Kanamori, H., Gavin, P.H., Hirshorn, B., Weinstein, S., 2011. Real-time W-Phase inversion during the 2011 off the Pacific coast of Tohoku earthquake. Earth Planets Space 63, 535–539.

Ebeling, C.W., Okal, E.A., 2012. An extension of the E/M0 tsunami earthquake discriminant T to regional distances. Geophys. J. Int. 190, 1640–1656.

Ebeling, C.W., Okal, E.A., Kalligeris, N., Synolakis, C.E., 2012. Modern seismological reassessment and tsunami simulation of historical Hellenic Arc earthquakes. Tectonophysics 530–531, 225–239.

El Alami, S.O., Tinti, S., 1991. A preliminary evaluation of the tsunami hazards in the Moroccan coasts. Sci.Tsunami Hazards 9, 31–38.
El Robrini, M., et al., 1985. Consequences of the El-asnam earthquakes: turbidity currents and slumps onthe Algerian Margin (Western Mediterranean). Geo-Mar. Lett. 5, 171–176.
Elias, A., Tapponnie, P., Singh, S.C., King, G.C.P., Briais, A., Daëron, M., Carton, H., Sursock, A., Jacques, E., Jomaa, R., Klinger, Y., 2007. Active thrusting offshore Mount Lebanon: source of the tsunamigenic AD 551 Beirut-Tripoli earthquake. Geology 35, 755–758.
Ellenblum, R., Marco, S., Agnon, A., Rockwell, T., Boas, A., 1998. Crusader castle torn apart by earthquake at dawn, 20 May 1202. Geology 26, 303–306.
Eredia, F., 1910. Il terremoto calabro messinese del 28 dicembre 1908. Annuario Scientifico ed Industriale, anno XLVI, 1909, Milano (in Italian).
Erginal, A.E., Kiyak, N.G., Özcan, H., 2009. Optically stimulated luminescence to date coastal dunes and a possitive tsunami layer on the Kavak Delta (Saros Gulf, NW Turkey). Turkish J. Earth Sci. 18, 465–474.
Eva, C., Rabinovich, A.B., 1997. The February 23, 1887 tsunami recorded on the Ligurian coast, western Mediterranean. Geophys. Res. Lett. 24, 2211–2214.
Evagelatou-Notara, F., 1986. The earthquake of 1402 in Achaia and in other regions. "Proc. Local Meeting Achaian Studies, Kalavryta, June 25–27, 1983", pp. 241–251 (in Greek).
Evagelatou-Notara, F., 1987/1988. Kai ta polla tis Peloponnisou seismou gegonasi paranaloma. Proceedings of the third Conference. Peloponnisian Studies, Kalamata, September 8–15, pp. 427–448 (in Greek).
Evagelatou-Notara, F., 1993. Seismoi sto Vyzantio apo ton 13o mexri ton 15o aiona, Istoriki exetasi (Earthquake in Byzantium from the 13th uo to 15th century, a historical examination). Parousia, 24, 1-179 (in Greek with Engl. Summary).
Faccenna, C., Piromallo, C., Crespo-Blanc, A., Jolivet, L., Rossetti, F., 2004. Lateral deformation and the origin of the western Mediterranean arcs. Tectonics 23, TC1012.
Favalli, M., Boschi, E., Mazzarini, F., Pareschi, M.T., 2009. Seismic and landslide source of the 1908 Straits of Messina tsunami (Sicily, Italy). Geophys. Res. Lett. 36, Art.N. L16304.
FEMA, 2000. Risk mapping assessment and planning. https://www.fema.gov/national-flood-insurance-program-flood-hazard-mapping (accessed 08.11.2015.).
Field, M.E., Gardner, J.V., 1990. Pliocene–Pleistocene growth of the Rio Ebro margin, northeast Spain: a prograding-slope model. Geol. Soc. Am. Bull. 102, 721–733.
Finkl, C.W., Pelinovsky, E., Cathcart, R.B., 2012. A review of potential tsunami impacts on the Suez Canal. J. Coastal Res. 28 (4), 745–759.
Flemming, N.C., Webb, C.O., 1986. Tectonic and eustatic coastal changes during the last 10,000 years derived from archaeological data. Zeitschrift für Geomorphologie N. F. 62, 1–29.
Flerit, F., Armijo, R., King, G.C.P., Meyer, B., Barka, A., 2003. Slip partitioning in the Sea of Marmara pull-apart determined from GPS velocity vectors. Geophys. J. Int. 154, 1–7.
Flood, S., Schechtman, J., 2012. The rise of resilience: evolution of a new concept in coastal planning in Ireland and the US. Ocean Coast Manag 102, 19–31.
Fokaefs, A., Papadopoulos, G.A., 2007. Tsunami hazard in the Eastern Mediterranean: strong earthquakes and tsunamis in Cyprus and the Levantine Sea. Nat. Hazards 40 (3), 503–526.
Folkman, Y., Mart, Y., 2008. Newly recognized eastern extension of the Nile deep-sea fan. Geology 36 (12), 939–942.
Fonseca, J.D., 2005. 1775-O Terramoto de Lisboa (The Lisbon Earthquake). 2ª edição, Argumentum (in English and Portuguese).
Fracassi, U., Di Bucci, D., Ridente, D., Trincardi, F., Valensise, G., 2012. Recasting historical earthquakes in coastal areas (Gargano Promontory, Italy): insights from marine paleoseismology. Bull. Seismol. Soc. Am. 102 (1), 1–17.
Friedrich, W., Kromer, L.B., Friedrich, M., Heinemeier, J., Pfeiffer, T., Talamo, S., 2006. Santorini eruption radiocarbon dated to 1627-1600 BC. Science 312, 548.

Fritzalas, C., Papadopoulos, G.A., 1988. Volcanic risks and urban planning in the region of Santorini volcano, South Aegean, Greece. In: Marinos, P., Koukis, G. (Eds.), Proceedings International Symposium on Enginigeering Geology as Related to the Study, Preservation & Protection of Ancient Works, Monuments & Historical Sites, Balkema, Athens, September 1988, vol. 3, pp. 1321–1327.

Frost, F.J., Hadjidaki, E., 1990. Excavations at the harbor of Phalasarna in Crete: the 1988 Season. Hesperia, 59, 513–527.

Fujii, Y., Satake, K., Sakai, S.-I., Shinohara, M., Kanazawa, T., 2011. Tsunami source of the 2011 off the Pacific coast of Tohoku Japan earthquake. Earth Planets Space 63, 815–820.

Fukao, Y., 1973. Thrust faulting at a lithospheric plate boundary The Portugal Earthquake of 1969. Earth Planet. Sci. Lett. 18, 205–216.

Gaki-Papanastassiou, K., Maroukian H., Papanastassiou, D., Palyvos, N., Lemmeille, F., 2001. Geomorphological study in the Lokrian Coast of N. Evoikos Gulf (Central Greece) and evidence of papaeoseismic destructions. Report du 36e Congres de la CIESM 36, 25.

Galanopoulos A.G., 1957a. Bulletin of the Seismological Institute, Year 1956. Natl. Observatory Athens, Athens.

Galanopoulos, A.G., 1957b. The seismic sea-wave of 9 July 1956. Praktika Acad. Athens 32, 90–101 (in Greek with Engl. abstr.).

Galanopoulos, A.G., 1960. Tsunamis observed on the coasts of Greece from Antiquity to present time. Ann. Geofis. 13, 369–386.

Galanopoulos, A.G., Bacon, E., 1970. Atlantis – the truth behind the legend. Nelson, United Kingdom, 216 pp.

Galanopoulos, A.G., Comninakis, P., Delibasis, N., 1964. A tsunami generated by an Earth slump set in motion without shock. Annales Géologique des Pays Helléniques 16, 93–110 (in Greek with Engl. Summary).

Galili, E., Horwitz, L.K., Hershkovitz, I., Eshed, V., Salamon, A., Weinstein-Evron, M., Greenfield, H., 2008. Comment on "Holocene tsunamis from Mount Etna and the fate of Israeli Neolithic communities" by Preschi, M.T., Boschi, E., Favalli, M. Geophys. Res. Lett. 35, L08311.

Garcia-Orellana, J., et al., 2006. Identifying instrumental and historical earthquake records in the SW Iberian Margin using 210Pb turbidite chronology. Geophys. Res. Lett. 33, L24601.

Gardi, A., Valencia, N., Guillande, R., André, C., 2011. Inventory of uncertainties associated with the process of tsunami damage assessment on buildings (SCHEMA FP6 EC co-funded project). Nat. Hazards Earth Syst. Sci. 11, 883–893.

Geist, E., Dmowska, R., 1999. Local tsunamis and distributed slip at the source. Pure Appl. Geophys. 154, 485–512.

Geist, E.L., Parsons, T., 2006. Probabilistic analysis of tsunami hazards. Nat. Hazards 37, 277–314.

Gennesseaux, M., Mauffret, A., Pautot, G., 1980. Les glissement sous-marins de la pente continentale niçoise et la rupture de câbles en mer Ligure (Méditerranee occidentale). Comptes Rendus de l'Académie des Sciences Paris 290, D 959.

Gerardi, F., Barbano, M.S., De Martini, P.M., Pantosti, D., 2008. Discrimination of tsunami sources (earthquake versus landslide) on the basis of historical data in eastern Sicily and southern Calabria. Bull. Seismol. Soc. Am. 98 (2), 2795–2805.

Gianfreda, F., Mastronuzzi, G., Sanso, P., 2001. Impact of historical tsunamis on a sandy coastal barrier: an example from the northern Gargano coast, southern Italy. Nat. Hazards Earth Syst. Sci. 1, 213–219.

Glimsdal, S., L'Heureux, J.-S., Harbitz, C., Løvholt, F., 2014. The landslide tsunami at Statland, mid-Norway, January 2014. Geophys. Res. Abstr., v. 17, EGU2015-11722, 2015, EGU General Assembly 2014.

Goff, J., Chagué-Goff, C., Nichol, S., Jaffe, B., Dominey-Howes, D., 2012. Progress in palaeotsunami research. Sediment. Geol. 243–244, 70–88.

Goff, J., Lane, E.M., Arnold, J., 2009. The tsunami geomorphology of coastal dunes. Nat. Hazards Earth Syst. Sci. 9, 847–854.
Goff, J., Terry, J.P., Chagué-Goff, C., Goto, K., 2014. What is a mega-tsunami? Mar. Geol. 358, 12–17.
Gokon, H., Koshimura, S., Imai, K., Matsuoka, M., Nameyaga, Y., Nishimura, Y., 2014. Developing fragility functions for the areas affected by the 2009 Samoa earthquake and tsunami. Nat. Hazards Earth Syst. Sci. 14, 3231–3241.
Goldsmith, V., Gilboa, M., 1985. Development of an Israeli tidal atlas and comparison with other Mediterranean tidal data. Israeli National Oceanography Institute Reports H8/85, p. 28.
Gomis, D., Monserrat, S., Tintoré, J., 1993. Pressure-forced seiches of large amplitude in inlets of the Balearic Islands. J. Geophys. Res. 98, 14 437–14 445.
Gonzalez, M., Medina, R., Olabarrieta, M., Otero, L., 2010. Tsunami hazard assessment on the Southern Coast of Spain. Turkish J. Earth Sci. 19 (3), 351–366.
Goodman-Tchernov, B.N., Dey, H.W., Reinhardt, E.G., McCoy, F., Mart, Y., 2009. Tsunami waves generated by the Santorini eruption reached Eastern Mediterranean shores. Geology 37, 943–946.
Goto C., 1990. Simulation models of timber and oil spread due to tsunamis, 2nd UJNR Tsunami Workshop, November 5–6, 1990, Honolulu, Hawaii, pp. 63–66.
Goto, K., Chagué-Goff, C., Fujino, S., Goff, J., Jaffe, B., Nishimura, Y., Richmond, B., Suguwara, D., Szczuciński, W., Tappin, D.R., Witter, R., Yulianto, E., 2011. Newinsights into tsunami risk from the 2011 Tohoku-oki event. Mar. Geol. 290, 46–50.
Gràcia, E., Bartolomé, R., Lo Iacono, C., Moreno, X., Stich, D., Martínez-Díaz, J.J., Bozzano, G., Martínez-Loriente, S., Perea, H., Masana, E., Dañobeitia, J.J., Tello, O., Sanz, J.L., Carrño, E., EVENT-SHELF team, 2012. Acoustic and seismic imaging of the active Adra Fault (NE Alboran Sea): in search for the source of the 1910 Adra Earthquake. Nat. Hazards Earth Syst. Sci. 12, 3255–3267.
Gràcia, E., Pallàs, R., Soto, J.I., Comas, M., Moreno, X., Masana, E., Santanach, P., Diez, S., García, M., Dañobeitia, J.J., 2006. Active faulting offshore SE Spain (Alboran Sea): implications for earthquake hazard assessment in the Southern Iberian Margin. Earth Planet. Sci. Lett. 241 (3–4), 734–749.
Gràcia, E., et al., 2003a. Mapping active faults offshore Portugal (36°N-38°N): implications for seismic hazard assessment along the southwest Iberian margin. Geology 31 (1), 83–86.
Gràcia, E., et al., 2003b. Crustal architecture and tectonic evolution of the Gulf of Cadiz (SW Iberian Margin) at the convergence of the Eurasian and African Plates. Tectonics 22 (4), 1033–1058.
Gràcia, E., et al., 2010. Holocene earthquake record offshore Portugal (SW Iberia): testing turbidite paleoseismology in a slow-convergence margin. Quat. Sci. Rev. 29, 1156–1172.
Grauert, M., Björck, Bondevik, S., 2001. Storegga tsunami deposits in a coastal lake on Suduroy, the Faroe Islands. Boreas 30, 263–271.
Graziani, L., Maramai, A., Tinti, S., 2006. A revision of the 1783-1784 Calabrian (southern Italy) tsunamis. Nat. Hazards Earth Syst. Sci. 6, 1053–1060.
Grezio, A., Sandri, L., Marzocchi, W., Argnani, A., Gasparini, P., Selva, J., 2012. Probabilistic tsunami hazard assessment for Messina Strait Area (Sicily, Italy). Nat. Hazards 64, 329–358.
Guidoboni, E., Comastri, A., 1997. The large earthquake of 8 August 1303 in Crete: seismic scenario and tsunami in the Mediterranean area. J. Seismol. 1, 55–72.
Guidoboni, E., Comastri, A., 2005. Catalogue of earthquakes and tsunamis in the Mediterranean area from the 11th to the 15th century. INGV-SGA, Rome-Bologna, Italy, 1037.
Guidoboni, E., Comastri, A., Traina, G., 1994. Catalogue of ancient earthquakes in the Mediterranean area up to the 10th Century. ING-SGA, Rome-Bologna, Italy, 504.
Guidoboni, E., Tinti, S., 1988. A review of the historical 1627 tsunami in the Southern Adriatic. Sci. Tsunami Hazards 6, 11–16.

Gusiakov, V., 2009. Tsunami history. In: Robinson, A., Bernard, E. (Eds.), The Sea, vol. 15. Harvard University Press, Cambridge, pp. 23–53.

Gutscher, M.A., Malod, J., Rehault, J.-P., Contrucci, I., Klingelhoefer, F., Mendes-Victor, L., Spakman, W., 2002. Evidence for active subduction beneath Gibraltar. Geology 30, 1071–1074.

Hadjidaki, E., 1988. Preliminary report of excavations at the harbor of Phalasarna in West Crete. Am. J. Archaeol. 92, 463–479.

Hadjidaki, E., 2001. The Roman destruction of Phalasarna. In: Archaeology of the Roman Empire–A tribute to the life and works of Professor Barri Jonea, BAR International Series, vol. 940, pp. 155–166.

Hadler, H., Vött, A., D., Koster, B., Mathes-Schmidt, M., Mattern, T., Ntageretzis, K., Reicherter, K., Willershäuser, T., 2013. Multiple late-Holocene tsunami landfall in the eastern Gulf of Corinth recorded in the palaeotsunami geo-archive at Lechaion, harbour of ancient Corinth (Peloponnese, Greece). Zeitschrift für Geomorphologie 57, (Suppl. 4), 139–180.

Hadler, H., Vött, A., Koster, B., Mathes-Schmidt, M., Mattern, T., Ntageretzis, K., Reicherter, K., Sakellariou, D., & Willershäuser, T., 2011. Lechaion, the ancient harbour of Corinth (Peloponnese, Greece) destroyed by tsunamigenic impact. 2nd INQUA-IGCP-567 Internat. Workshop on Active Tectonics, Earthquake Geology, Archaeology & Engineering, Corinth, Greece.

Hammitzsch, M., Spazier, J., Reißland. S, Necmioglu, O., Comoglu, M., Sozdinler, O.C., Carrilho, F., Wächter, J., 2015. TRIDEC Cloud – aWeb-based Platform for Tsunami EarlyWarning tested with NEAMWave14 Scenarios. Geophys. Res. Abstr., 17, EGU2015-9084, EGU General Assembly 2015.

Hancilar, U., 2012. Identification of elements at risk for a credible tsunami event for Istanbul. Nat. Hazards Earth Syst. Sci. 12, 107–119.

Hara, T., 2011. Magnitude determination using duration of high frequency energy radiation and displacement amplitude: application to the 2011 off the Pacific coast of Tohoku earthquake. Earth Planets Space 63, 525–528.

Harada, K., Imamura, F., 2005. Effects of coastal forest on tsunami hazard mitigation – a preliminary investigation. In: Satake, K. (Ed.), Tsunamis: Case Studies and Recent Developments. Springer, The Netherlands, pp. 279–292.

Harbitz, C.B., 1992. Model simulations of tsunamis generated by the Storegga slides. Mar. Geol. 105, 1–21.

Harbitz, C.B., Glimsdal, S., Løvholt, F., Kveldsvik, V., Pedersen, G.K., Jensen, A., 2014. Rockslide tsunamis in complex fjords: from an unstable rock slope at Åkerneset to tsunami risk in western Norway. Coast. Eng. 88, 101–122.

Hasegawa, A., Yoshida, K., Okada, T., 2011. Nearly complete stress drop in the 2011 Mw9.0 off the Pacific coast of Tohoku earthquake. Earth Planets Space 63, 703–707.

Haslett, S.K., Bryant, E.A., 2008. Historic tsunami in Britain since AD 1000: a review. Nat. Hazards Earth Syst. Sci. 8, 587–601.

Hatori, T., 1986. Classification of tsunami magnitude scale. Bull. Earthq. Res. Inst. Univ. Tokyo 61, 503–515 (in Japanese with Engl. abstr.).

Heiken, G.H., McCoy, F.W., 1984. Caldera development during the Minoan eruption, Thera, Cyclades, Greece. J. Geophys. Res. 89, 8441–8462.

HELZEL Messtechnik GmbH, 2011. WERA Ocean radar in Chile observed tsunami signatures after the earthquake in Japan on March 11, 2011. Press Release, May 2011, pp. 1–2.

Hieke, W., 1984. A thick Holocene homogenite from the Ionian Abyssal Plain (Eastern Mediterranean). Mar. Geol. 55, 63–78.

Hill, J., Collins, G., Smith, R., Piggott, M., 2014. Numerical modelling of the Storegga tsunami: consequences to the UK. Geophys. Res. Abstr., 16, EGU2014-4052, 2014, EGU General Assembly.

Hills, J.G., Mader, C.L., 1997. Tsunami produced by the impacts of small asteroids. Ann. New York Acad. Sci. 822, 381–394.
Hinata, H., Fujii, S., Furukawa, K., et al., 2011. Propagating tsunami wave and subsequent resonant response signals detected by HF radar in the Kii Channel, Japan. Estuarine. Coastal Shelf Sci. 95 (1), 268–273.
Hindson, R.A., Andrade, C., Dawson, A.G., 1996. Sedimentary processes associated with the tsunami generated by the 1755 Lisbon Earthquake on the Algarve Coast, Portugal. Phys. Chem. Earth 21, 57–63.
Horsburgh, K.J., Wilson, C., Baptie, B.J., Cooper, A., Cresswell, D., et al., 2008. Impact of a Lisbon-type tsunami on the U.K. coastline and the implications for tsunami propagation over broad continental shelves. J. Geophys. Res., 113, C04007.
Hoshiba, M., Iwakiri, K., Hayashimoto, N., et al., 2011. Outline of the 2011 off the Pacific coast of Tohoku earthquake (Mw9.0)-earthquake early warning and observed seismic intensity. Earth Planets Space 63, 547–551.
Ide, S., Baltay, A., Beroza, G.C., 2011. Shallow dynamic overshoot and energetic rupture in the 2011 Mw9.0 Tohoku-Oki earthquake. Science 332, 1426–1429.
Iglesias, O., Lastras, G., Canals, M., Olabarrieta, M., González, M., Aniel-Quiroga, I., Otero, L., Durán, R., Amblas, D., Casamor, J.L., Tahchi, E., Tinti, S., De Mol, B., 2012. The BIG'95 submarine landslide-generated tsunami: a numerical simulation. J. Geol. 120 (1), 31–48.
Iida, K., 1956. Earthquakes accompanied by tsunamis occurring under the sea off the islands of Japan. J. Earth Sci. Nagoya Univ. 4, 1–43.
Iida, K., 1970. The generation of tsunamis and the focal mechanism of earthquakes. In: Adams, W.M. (Ed.), Tsunamis in the Pacific Ocean. East-West Center Press, Honolulu, pp. 3–18.
Iida, K., Cox, D.C., Pararas-Carayannis, G., 1967. Preliminary Catalog of Tsunamis Occurring in the Pacific Ocean, Data Rep. 5, HIG-67-10. Hawaii Inst. of Geophys., Univ. of Hawaii, Hawaii.
Iinuma, T., Ohzono, M., Ohta, Y., Miura, S., 2011. Coseismic slip distribution of the 2011 off the Pacific coast of Tohoku earthquake (M9.0) estimated based on GPS data-Was the asperity in Miyagi-oki ruptured? Earth Planets Space 63, 643–648.
Imai, K., Matsutomi, H., 2005. Fluid Force on Vegetation due to tsunami flow on a sand spit. In: Satake, K. (Ed.), Tsunamis: Case Studies and Recent Developments. Springer, Berlin, pp. 293–304.
Imamura, A., 1942. History of Japanese tsunamis. Kayo-No-Kagaku (Oceanography) 2, 74–80, (in Japanese).
Imamura, A., 1949. List of tsunamis in Japan. J. Seismol. Soc. Japan 2, 23–28, (in Japanese).
IOC-Intergovernmental Oceanographic Commission, 1999. Historical tsunami database for the Pacific, 47BC-1998AD. Tsunami Laboratory, Siberian Russian Academy of Sciences, Novosibirsk, Russia. http://tsun.sscc.ru/HTDBPac1/.
IOC-Intergovernmental Oceanographic Commission, 2013. Tsunami Glossary, Technical Series 85, UNESCO, 1-42.
Ionescu, R., Partheniu, R., Cioflan, C., et al., 2014. Tsunami early warning system for the western coast of the Black Sea. Geophys. Res. Abstr., v.16, EGU2014-3569, EGU General Assembly 2014.
Iwan, W.D. (Ed.), 2006. The Great Sumatra Earthquakes and Indian Ocean Tsunamis of 26 December 2004 and 28 March 2005 Reconnaissance Report. Earthquake Spectra, Special Issue III, vol. 22. Publication Number 2006-05, S1-S900.
Jacques, F., Bousquet, B., 1984. Le raz de mare du 21 juillet 365. Du cataclysme local à la catastrophe cosmique. Mélanges de l' École Française de Rome, Antiquité 96, pp. 423–461.
Jørstad, F.A., 1968. Waves generated by landslides in Norwegian fjords and lakes. Norwegian Geotechnical Institute, Oslo, Publ. nr. 79, 13–32.

Kamigaichi, O, 2012. JMA's response to Off the Pacific Coast of Tohoku earthquake and planned improvements of tsunami warning. Japan–IOC/UNESCO/UNU Symposium "The Great East Japan Tsunami on 11 March 2011 and Tsunami Warning Systems: Policy Perspectives", Feb 16–17, 2012. UN University, Tokyo, Japan. Available from: http://ioc-tsunami.org/index.php?option=com_oe&task=view Document Record&docID=8619.

Kanamori, H., 1972. Mechanism of tsunami earthquakes. Phys. Earth Planet. Inter. 6, 346–359.

Kanamori, H., 1977. The energy release in great earthquakes. J. Geophys. Res. 82, 2981–2987.

Kanamori, H., 2006. Seismological aspects of the December 2004 Great Sumatra-Andaman Earthquake. In: Iwan, W.D. (Ed.), The Great Sumatra Earthquakes and Indian Ocean Tsunamis of 26 December 2004 and 28 March 2005 Reconnaissance Report. Earthquake Spectra, Special Issue III, vol. 22, Publication Number 2006-05, S1-S11.

Kanamori, H., Rivera, L., 2008. Source inversion of W phase: speeding up seismic tsunami warning. Geophys. J. Internat. 175, 222–238.

Karagiannis, G.M., Saini, K., Synolakis, C., 2015. Lessons from the first EU tsunami simulation exercise. Presented at "The International Emergency Management Society USA Conference – Global Response for Capacity Building of Disaster Preparedness, At Hattiesburg, Mississippi, USA, July 2014." Available from: http://www.researchgate.net/publication/269410751.

Kastens, K.A., Cita, M.B., 1981. Tsunami-induced transport in the abyssal Mediterranean Sea. Geol. Soc. Am. Bull. 92, 845–857.

Kawamata, S., Imamura, F., Shuto, N., 1993. Numerical Simulation of the 1883 Krakatau Tsunami. Intern. Assoc. Hydraulic Res. 1–4, Tokyo, 24–31.

Kawamura, K., Sasaki, T., Kanamatsu, T., Sakaguchi, A., Ogawa, Y., 2012. Large submarine landslides in the Japan Trench: a new scenario fora additional tsunami generation. Geophys. Res. Lett. 39, L05308.

Kharif, Ch., Pelinovsky, E., 2005. Asteroid impact tsunamis. Comptes Rendus Physique 6, 361–366.

Kijko, A., 2004. Estimation of the maximum earthquake magnitude Mmax. Pure Appl. Geophys 161, 1–27.

Kijko, A., Sellevoll, M.A., 1989. Estimation of earthquake hazard parameters from incomplete data files: Part I, Utilization of extreme and complete catalogues with different threshold magnitudes. Bull. Seism. Soc. Am. 79, 645–654.

Kijko, A., Sellevoll, M.A., 1992. Estimation of earthquake hazard parameters from incomplete data files, Part II, Incorporation of magnitude heterogeneity. Bull. Seism. Soc. Am. 82, 120–134.

Kijko, A., Singh, M., 2011. Statistical tools for maximum possible earthquake magnitude estimation. Acta Geophysica 59, 674–700.

Kijko, A., Smit, A., 2012. Extension of the Aki-Utsu b-value Estimator for Incomplete Catalogs. Bull. Seism. Soc. Am, June 2012 102, 3.

Kontoes, Ch., Herekakis, Th., Ieronymidi, E., Keramitsoglou, I., Fokaefs, A., Papadopoulos, G.A., Paralikidis, S., Aifantopoulou, D., Deflorio, A.M., Iasillo, D., Kiranoudis, Ch.T., 2012. Mapping seismic vulnerability and risk of cities: The MASSIVE Project. J. Earth Sci. Eng. 2, 496–513.

Kontopoulos, N., Avramidis, P., 2003. A late Holocene record of environmental changes from the Aliki lagoon, Egion, North Peloponnesus, Greece. Quat. Internat 111, 75–90.

Kortekaas, S., Papadopoulos, G.A., Ganas, A., Cundy, A.B., Diakantoni, A., 2011. Geological identification of historical tsunamis in the Gulf of Corinth, Central Greece. Nat. Hazards Earth Syst. Sci. 11, 2029–2041.

Kortekaas, S., Papadopoulos, G.A., Ganas, A., Diakantoni, A., 2003. Geological identification of historical tsunamis in the Gulf of Corinth, Greece. Geophys. Res. Abs. 5 09909, 2003.

Koshiishi, H., 1998. Hazards to civilization from the collision of minor bodies. In:Yabushita, S., Henrand, J. (Eds.), Dynamics of Comets and Asteroids and Their Role in Earth History. Kluwer Academic, Dordrecht, pp. 267–281.

Koshimura, S., 2012.The impact of the 2011 Tohoku earthquake tsunami disaster. Lesson towards tsunami-resilient communities. Japan–IOC/UNESCO/UNU Symposium "The Great East Japan Tsunami on 11 March 2011 and Tsunami Warning Systems: Policy Perspectives", Feb 16–17, 2012. UN University, Tokyo, Japan. Available from: http://ioc-tsunami.org/index.php?option=com_oe&task=view Document Record&docID=8608.

Koshimura, S., Namegaya, Y., Yanagisawa, H., 2009b. Tsunami Fragility: a new measure to assess tsunami damage. J. Disaster Res. 4, 479–488.

Koshimura, S., Oie, T., Yanagisawa, H., Imamura, F., 2009a. Developing fragility curves for tsunami damage estimation using numerical model and post-tsunami data from Banda Aceh, Indonesia. Coast. Eng. J. 51, 243–273.

Koutitas, C.G., Papadopoulos, G.A., 1998. Numerical simulation of the aseismically induced tsunami of 7 February 1963 in the Western Corinthos Bay. In: Proc. Internat. Tsunami Confer. Paris. pp. 247–254, May 26–28, 1998.

Koutitas, Ch., Karambas, Th., 2005. A computational approach to design codes for tsunami-resisting coastal structures. ISET Journal of Earthquake Technology, No 461 42 (4), 137–145.

Kraft, J.C., Rapp, G.R., Gifford, J.A., 2005. Coastal change and archaeological settings in Elis. Hesperia 74 (1), 1–39.

Kulkarni, R., Wong, I., Zachariasen, J., 2013. Statistical analyses of great earthquake recurrence along the Cascadia Subduction Zone. Bull. Seism. Soc. Am. 103, 3205–3221.

La Rocca, M., Galluzzo, D., Saccorotti, G., Tinti, S., Cimini, G.B., Del Pezzo, E., 2004. Seismic Signals Associated with Landslides and with a Tsunami at Stromboli Volcano, Italy. Bull. Seismol. Soc. Am. 94, 1850–1867.

Lambotte, S., Rivera, L., Hinderer, J., 2006. Rupture length and duration of the 2004 Aceh-Sumatra earthquake from the phases of the Earth's gravest free oscillations. Geophys. Res. Lett. 33, L03307.

Lario, J., Luque, L., Zazo, C., Goy, J.L., Spencer, C., Cabero, A., Bardají, T., Borja, F., Dabrio, C.J., Civis, J., González-Delgado, J.A., Borja, C., Alonso-Azcárate, J., 2010. Tsunami vs. Storm surge deposits: a review of the sedimentological and geomorphological records of extreme wave events (EWE) during the Holocene in the gulf of Cadiz. Spain. Zeitschrift für Geomorphologie 54 (Suppl.3), 301e316.

Lario, J., Zazo, C., Goy, J.L., Silva, P.G., Bardaji, T., Cabero, A., Dabrio, C.J., 2011. Holocene palaeotsunami catalogue of SW Iberia. Quat. Internat. 242, 196–200.

Larroque, C., Scotti, O., Mansour, I., 2012. Reappraisal of the 1887 Ligurian earthquake (western Mediterranean) from macroseismicity, active tectonics and tsunami modelling. Geophys. J. Int. 190, 87–104.

Lastras, G., et al., 2002. Seafloor imagery from the BIG'95 debris flow, western Mediterranean. Geology 10 (30), 871–874.

Lastras, G., et al., 2004. Shallow slides and pockmark swarms in the Eivissa Channel, Western Mediterranean Sea. Sedimentology 51, 1–14.

Lastras, G., et al., 2007. Slope instability along the northeastern Iberian and Balearic continental margins. Geologica Acta 5, 35–47.

Latter, J.H., 1981. Tsunamis of volcanic origin: summary of causes, with particular reference to Krakatoa, 1883. Bull. Volcanol. 44 (3), 467–489.

Lauterjung, J., Münch, U., Rundloff, A., 2010. The challenge of installing a tsunami early warning system in the vicinity of the Sunda Arc, Indonesia. Nat. Hazard Earth Syst. 10, 641–646.

Lay, T., Kanamori, H., Ammon, C.J., et al., 2006. The great Sumatra-Andaman earthquake of 26 Dcember 2004. Science 308 (5725), 1127–1133.

Lay, Th., Ammon, Ch.J., Kanamori, H., Xue, L., Kim, M.J., 2011. Possible large near-trench slip during the great the 2011 Tohoku (Mw9.0) earthquake. Earth Planets Space 63, 687–692.

Leone, F., Lavigne, F., Paris, R., Denain, J.C., Vinet, F., 2011. A spatial analysis of the December 26th, 2004 tsunami-induced damages: lessons learned for a better risk assessment integrating buildings vulnerability. Appl. Geogr. 31, 363–375.

Levin, B., Nosov, M., 2009. Physics of Tsunamis. Springer, Berlin, 327 pp.

Liu, J.-Y, Sun, Y.-Y., 2011. Seismo-traveling ionospheric distrurbances of ionograms observed during the 11 March 2011 M9.0 Tohoku earthquake. Earth Planets Space 63, 897–902.

Lo Iacono, C., Gràcia, E., Zaniboni, F., Pagnoni, G., Tinti, S., Bartolomé, R., Masson, D., Wynn, R., Lourenço, N., Pinto de Abreu, M., Dañobeitia, J.J., Zitellini, N., 2012. Large, deep water slope failures: implications for landslide-generated tsunamis. Geology 40 (10), 931–934.

Lockridge, 1988. Historical tsunamis in the Pacific basin. In: El-Sabh, M.I., Murty, T.S. (Eds.), Natural and Man-made Hazards. Kluwer, Reidel, Dordrecht, pp. 171–181.

Lomax, A., Michelini, A., 2011. Tsunami early warning using earthquake rupture duration and P-wave dominant period: the importance of length and depth of faulting. Geophys. J. Int. 185, 283–291.

Lomax, A., Michelini, A., 2012. Tsunami early warning within 5 minutes. Pure and Appl. Geophys. 170, 1385–1395.

Lomax, A., Michelini, A., Bernardi, F., Lauciani, V., 2015. Rapid and robust characterization of the earthquake source for tsunami early-warning. Geophys. Res. Abstr. 17, EGU2015-13568.

Lomax, A., Michelini, A., Piatanesi, A., 2007. An energy-duration procedure for rapid determination of earthquake magnitude and tsunamigenic potential. Geophys. J. Int. 170, 1195–1209.

Lorito, S., Selva, J., Basili, R., Romano, F., Tiberti, M.M., Piatanesi, A., 2015. Probabilistic hazard for seismically induced tsunamis: accuracy and feasibility of inundation maps. Geophys. J. Int. 200 (1), 574–588.

Lorito, S., Tiberti, M.M., Basili, R., Piatanesi, A., Valensise, G., 2008. Earthquake-generated tsunamis in the Mediterranean Sea: scenarios of potential threats to southern Italy. J. Geophys. Res. 113, B01301.

Løvholt, F., et al., 2009. Tsunami hazard studies in the eastern Hellenic Arc and Balearic Islands. EGU Gen. Ass., Vienna., April 19–24, 2009, Geophys. Res. Abstr. 11, Abstract EGU2009-2728-4.

Løvholt, F., Glimsdal, S., Lynett, P., Pedersen, G., 2015. Simulating tsunami propagation in fjords with long-wave models. Nat. Hazards Earth Syst. Sci. 15, 657–669.

Luis, J.F., 2007. Mirone: a multi-purpose tool for exploring grid data. Comput. Geosci. 33, 31–41.

Luque, L., Lario, J., Silva, P.G., Zazo, G., Goy, J.L., Dabrio, C.J., 2002. Sedimentary record of a tsunami during Roman times, Bay of Cadiz, Spain. J. Quat. Sci. 17 (5–6), 623–631.

Lykousis, V., Roussakis, G., Sakellariou, D., 2008. Slope failures and stability analysis of shallow water prodeltas in the active margins of Western Greece, northeastern Mediterranean Sea. Internat. J. Earth Sci., 98, 807–822.

Macias, J., Castro, M., Vida, J.M. G., de la Asuncion, M., Ortega, S., 2014. HySEA: an operational GPU-based model for Tsunami Early Warning Systems. Geophys. Res. Abstr., 16, EGU2014-14217, 2014, EGU General Assembly 2014.

Maeda, T., Furumura, T., Sakai, S.-I., Shinohara, M., 2011. Significant tsunami observed at the ocean-bottom pressure gauges at 2011 off the Pacific coast of Tohoku earthquake. Earth Plan. Space 63, 803–808.

Makris, J., Papoulia, J., Tsambas, A., 2014. A real-time seismic and tsunami network in the Kyparissiakos Gulf, Greece. Bollettino di Geofisica Teorica ed Applicata 55, 561–587.

Maldonado, A., Stanley, D.J., 1976. The Nile Cone: submarine fan development by cyclic sedimentation. Mar. Geol. 20 (1), 27–35.

Mamo, B., Strotz, L., Dominey-Howes, D., 2009. Tsunami sediments and their foraminiferal assemblages. Earth Sci. Rev. 96, 263–278.

Maramai, A., Brizuela, B., Graziani, Laura, 2014. The Euro-Mediterranean tsunami catalogue. Ann. Geophys. 57, S0435.

Maramai, A., Graziani, L., Alessio, G., Burrato, P., Colini, L., Cucci, L., Nappi, R., Nardi, A., Vilardo, G., 2005b. Near- and far-field survey report of the 30 December 2002 Stromboli (Southern Italy) tsunami. Mar. Geol. 215, 93–106.

Maramai, A., Graziani, L., Tinti, S., 2005a. Tsunamis in the Aeolian islands (southern Italy): a review. Mar. Geol. 215, 11–21.

Maramai, A., Graziani, L., Tinti, S., 2007. Investigation on tsunami effects in the central Adriatic Sea during the last century – a contribution. Nat. Hazards Earth Syst. Sci. 7, 15–19.

Marco, S., Agnon, A., Ellenblum, R., Eidelman, A., Basson, U., Boas, A., 1997. 817-year-old walls offset sinistrally 2.1 m by the Dead Sea Transform, Israel. J. Geodyn. 24, 11–20.

Marco, S., Rockwell, T.K., Heimann, A., Frieslander, U., Agnon, A., 2005. Late Holocene slip of the Dead Sea Transform revealed in 3D palaeoseismic trenches on the Jordan Gorge segment. Earth Planet. Sci. Lett. 234, 189–205.

Marcos, M., Monserrat, S., Medina, R., Vidal, C., 2003. Influence of the atmospheric wave velocity in the coastal amplification of meteotsunamis. In: Yalciner, A.C., Pelinovsky, E., Okal, E., Synolakis, C.E. (Eds.), Submarine Landlsides and Tsunamis. Kluwer Academic, Dordrecht, pp. 243–249.

Martinez-Loriente, S., Gràcia, E., Bartolomé, R., et al., 2013. Active deformation in old oceanic lithosphere and significance for earthquake hazard: seismic imaging of the Coral Patch Ridge area and neighbouring abyssal plains (SW Iberian Margin). Geochem. Geophys. Geosyst. 14 (7), 2206–2231.

Martínez-Loriente, S., Gràcia, E., Bartolome, R., et al., 2013. Active deformation in old oceanic lithosphere and significance for earthquake hazard: seismic imaging of the Coral Patch Ridge area and neighboring abyssal plains (SW Iberian Margin). Geochem. Geophys. Geosyst. 14 (7), 2206–2231.

Maruyama, T., Tsugawa, T., Kato, H., et al., 2011. Ionospheric multiple stratifications and irregularities induced by the 2011 Tohoku earthquake. Earth Planets Space 63, 869–873.

Mas, E., Koshimura, S., Suppasri, A., Matsuoka, M., Matsuyama, M., Yoshii, T., Jimenez, C., Yamazaki, F., Imamura, F., 2012. Developing tsunami fragility curves using remote sensing and survey data of the 2010 Chilean tsunami in Dichato. Nat. Hazards Earth Syst. Sci. 12, 2689–2697.

Mascle, J., Mascle, G., 2012. Geological and Morpho-tectonic Map of the Mediterranean domain: a synthesis at the scale of 1:4000000. Spec. Public. CCGM/UNESCO, Paris.

Mastronuzzi, G., Pignatelli, C., Sansò, P., 2006. Boulder fields: a valuable morphological indicator of Paleotsunami in the Mediterranean Sea. Zeitschrift für Geomorphologie NF Suppl. -Bd. 146, 173–194.

Mastronuzzi, G., Pignatelli, C., Sansò, P., Selleri, G., 2007. Boulder accumulations produced by the 20th February 1743 tsunami along the coast of southeastern Salento (Apulia region, Italy). Mar. Geol. 242 (1), 191–205.

Mastronuzzi, G., Sansò, P., 2000. Boulders transport by catastrophic waves along the Ionian coast of Apulia (southern Italy). Mar. Geol. 170, 93–103.

Mastronuzzi, G., Sansò, P., 2004. Large Boulder Accumulations by Extreme Waves along the Adriatic Coast of Southern Apulia (Italy). Quat. Int. 120, 173–184.

Mathes-Schmidt, M., Schwarzbauer, J., Papanikolaou, I., Syberberg, F., Thiele, A., Wittkopp, F., Reicherter, K., 2013. Geochemical and micropaleontological investigation of tsunamigenic layers along the Thracian Coast (Northern Aegean Sea, Greece). Zeitschrift für Geomorphologie 57 (Suppl. 4), 005–027.

Matias, L.M., Cunha, T., Annunziato, A., Baptista, M.A., Carrilho, F., 2013. Tsunamigenic earthquakes in the Gulf of Cadiz: fault model and recurrence. Nat. Hazards Earth Syst. Sci. 13, 1–13.

Matsutomi H., 1991. Impulsive force of timbers drifted by tsunamis. In: Brennan, A.M., Lander, J.F. (Eds.), 2nd UJNR Tsunami Workshop, November 5–6, 1990, Honolulu, Hawaii.

McCoy, F.W., Heiken, G., 2000. Tsunami generated by the Late Bronze Age eruption in Thera (Santorini), Greece. Pure Appl. Geophys. 157, 1227–1256.

McCoy, F.W., Papadopoulos, G.A., 2001. Tsunami generation during the Late Bronze Age Eruption ofThera: evidence from tsunami deposits on Thera, Crete, West Turkey and the Deep Sea. Am. J. Archaeol. 105, 258–259 (abstr.).

Meghraoui, M., Maouche, S., Chemaa, B., Cakir, Z., Aoudia, A., Harbi, A., Alasset, P.J., Ayadi, A., Bouhadad, Y., Benhamouda, F., 2004. Coastal uplift and thrust faulting associated with the Mw = 6.8 Zemmouri (Algeria) earthquake of 21 May, 2003. Geophys. Res. Lett. 31, L19605.

Minoura, K., Imamura, F., Kuran, U., Nakamura, T., Papadopoulos, G.A., Sugawara, D., Takahashi, T., Yalçiner, A., 2005. A tsunami generated by a possible submarine slide: evidence for slope failure triggered by the North Anatolian Fault movement. Nat. Hazards 36, 297–306.

Minoura, K., Imamura, F., Kuran, U., Nakamura, T., Papadopoulos, G.A., Takahashi, T., Yalçiner, A., 2000. Discovery of Minoan tsunami deposits. Geology 28, 59–62.

Minoura, K., Imamura, F., Kuran, U., Nakamura, T., Papadopoulos, G.A., Takahashi, T., Yalciner, A.C., 2003. Tsunami hazards associated with explosion-collapse processes of a dome complex on Minoan Thera. Adv. Res. Workshop Underwater Ground Failures on Tsunami Generation, Modeling, Risk & Mitigation, Kluwer Academic, Dordrecht, pp. 229–236.

Minoura, K., Imamura, F., Sugawara, D., Kono, Y., Iwashita, T., 2001. The 869 Jögan tsunami deposit and recurrence interval of large-scale tsunami on the Pacific coast of northeast Japan. J. Nat. Dis. Sci. 23, 83–88.

Mitsoudis, D.A., Flouri, E.T., Chrysoulakis, N., et al., 2012. Tsunami hazard in the southeast Aegean Sea. Coast. Eng. 60, 136–148.

Monserrat, S., Ibbetson, A., Thorpe, A.J., 1991. Atmospheric gravity waves and the Rissaga phenomenon. Q. J. R. Meteor. Soc. 117, 553–570.

Monserrat, S., Vilibić, I., Rabinovich, A.B., 2006. Meteotsunamis: atmospherically induced destructive ocean waves in the tsunami frequency band. Nat. Hazards Earth Syst. Sci. 6, 1035–1051.

Morhange, C., Marriner, N., Pirazzoli, A., 2006. Evidence of late-Holocene tsunami events in Lebanon. Zeitschrift für Geomorphologie Supplement 146, 81–95.

Mori, N., Takahashi, T., Yasuda, T., Yanagisawa, H., 2011. Survey of the 2011 Tohoku earthquake tsunami inundation and run-up. Geophys. Res. Lett. 38, L010.

Mörner, N.-A., 2005. Active faults and paleoseismicity in Fennoscandia, especially Sweden. Primary structures and secondary effects. Tectonophysics 380, 139–157.

Morrisson, C., 2007. Le Monde Byzantin: I – L'Empire romain d' orient (330-641). Athens: Polis. (transl. in Greek).

Morton, R.A., Gelfenbaum, G., Jaffe, B.E., 2007. Physical criteria for distinguishing sandy tsunami and storm deposits using modern examples. Sediment. Geol. 200, 184–207.

Mourtzas, N.D., Kissas, C., Kolaiti, E., 2013. Archaeological and geomorphological indicators of the historical sea level changes and the related palaeogeographical reconstruction of the ancient foreharbour of Lechaion, East Corinth Gulf (Greece). Quat. Int. 332, 151–171.

Mourtzas, N.D., Marinos, P.G., 1994. Upper Holocene sea-level changes. Palaeogeographic evolution and its impact on coastal archaeological sites and monuments. Environ. Geol. 23, 1–13.

Müller, L., 1964. The rock slide in the Vajont Valley. Felsmechanik u. Ingenieuregeol., II/3-4, 148–212.
Murata, S., Imamura, F., Katoh, K., et al., 2010. Tsunami: To Survive From Tsunami. World Scientific, Singapore, 302 pp., Adv. Ser. on Ocean Engineering, 32.
Murty, T.S., 1977. Seismic sea waves – tsunamis. Bull. Fisheries Res. Board Canada 198, 1–337.
Murty, T.S., Aswathanarayana, U., Nirupama, N. (Eds.), 2007. The Indian Ocean Tsunami. Taylor & Francis, London, UK, 491 pp.
Murty, T.S., Loomis, H.G., 1980. A new objective tsunami magnitude scale Marine Geodesy 4, 267–282.
Nemer, T., Gomez, F., Al Haddad, S., Tabet, C., 2008. Coseismic growth of sedimentary basins along the Yammouneh strike-slip fault (Lebanon). Geophys. J. Int. 175, 1023–1039.
Neri, G., Barberi, G., Oliva, G., Orecchio, B., 2004. Tectonic stress and seismogenic faulting in the area of the 1908 Messina earthquake, south Italy. Geophys. Res. Lett. 31, L10602.
Nettles, M., Ekström, G., Koss, H., 2011. Centroid-moment-tensor analysis of the 2011 Tohoku earthquake and its larger foreshock and aftershocks. Earth Planets Space 63, 519–523.
Newman, A.V., Okal, E.A., 1998. Teleseismic estimates of radiated seismic energy: the E/Mo discriminant for tsunami earthquakes. J. Geophys. Res. 103 (B11), 26885–26898.
Nichol, S.L., Goff, J.R., Devoy, R.J., Hayward, B., Chague-Goff, C., James, I., 2007. Lagoon subsidence and tsunami on the West Coast of New Zealand. Sediment. Geol. 200, 248–262.
Ninkovich, D., Hays, J.D., 1972. Mediterranean Islands Arcs and Origin of High Potash Volcanoes. Earth Planet. Sci. Lett. 16, 331–345.
Noller, S., Panayides, I., Zomeni, Z., 2005. Report on the preliminary assessment of tsunami hazard in Cyprus, Lefkosia. Cyprus, p. 27.
Nomikou, P., Carey, S., Bell, K.L.C., et al., 2012. Tsunami hazard risk of a future volcanic eruption of Kolumbo submarine volcano. NE of Santorini Caldera, Natural Hazards, Greece.
Novikova, T., Papadopoulos, G.A., McCoy, F.W., 2011. Modeling of tsunami generated by the Giant Late Bronze Age eruption of Thera, South Aegean Sea, Greece. Geophys. J. Int. 186, 665–680.
Okada, Y., 1992. Internal deformation due to shear and tensile faults in a half-space. Bull. Seismol. Soc. Am. 82, 1018–1040.
Okal, E., 1988. Seismic parameters controlling far-field tsunami amplitudes: a review. Nat. Hazards 1, 67–96.
Okal, E.A., Synolakis, C.E., 2004. Source discriminants for near-field tsunamis. Geophys. J. Int. 158, 899–912.
Okal, E.A., Synolakis, C.E., Uslu, B., Kalligeris, N., Voukouvalas, E., 2009. The 1956 earthquake and tsunami in Amorgos, Greece. Geophys. J. Int. 178, 1533–1554.
Okal, E.A., Talandier, J., 1988. M_m: a variable-period mantle magnitude. J. Geophys. Res. 94, 4169–4193.
Olabarrieta, M., Medina, R., González, M., Otero, L., 2011. A finite volume-finite difference hybrid model for tsunami propagation and runup. Comput. Geosci. 37, 1003–1014.
Oliveira, C.S., 2008. Review of the 1755 Lisbon earthquake based on recent analyses of historical observations. In: Fréchet, J., et al. (Eds.), Historical Seismology – Interdisciplinary Studies of Past and Recent Earthquakes. Modern Approaches in Solid Earth Sciences vol. 2, pp. 262–300.
Omira, R., Baptista, M.-A., Leone, F., Matias, L., Mellas, S., Zourarah, B., Miranda, J.M., Carrilho, F., Cherel, J.-P., 2013. Performance of coastal sea-defense infrastructure at El Jadida (Morocco) against tsunami threat: lessons learned from the Japanese 11 March 2011 tsunami. Nat. Hazards Earth Syst. Sci. 13, 1779–1794.

Omira, R., Baptista, M.A., Miranda, J.M., Matias, L., Carrilho, L., Catita, C., Toto, E., Catita, C., 2009. Design of a sea-level tsunami detection network for the Gulf of Cadiz. Nat. Hazards Earth Syst. Sci. 9, 1327–1338.

Omira, R., Baptista, M.A., Miranda, J.M., Toto, E., Catita, C., Catalão, J., 2010. Tsunami vulnerability assessment of Casablanca Morocco using numerical modelling and GIS tools. Nat. Hazards Earth Syst. Sci. 54, 75–95.

Omori, F., 1910. Preliminary report on the Messina-Reggio earthquake of Dec. 28, 1908. Bull. Imperial Earthquake Investigation Committee 3(2), 37–45, Tokyo.

Onat, Y., Yalciner, A.C., 2013. Initial stage of database development for tsunami warning system along Turkish coasts. Ocean Eng. 74, 141–154.

Ozel, N.M., Necmioglu, O., Yalciner, A.C., et al., 2011. Tsunami hazard in the Easterm Mediterranean and its connected seas: towards a tsunami warning center in Turkey. Soil Dyn. Earthq. Eng. 31, 598–610.

Panza, G.F., Romanelli, F., Yanovskaya, T.B., 2000. Synthetic tsunami mareograms for realistic oceanic models. Geophys. J. Int. 141, 498–508.

Papadopoulos, G.A., Satake, K. (Eds.), 2005. Proceedings of the 22nd International Tsunami Symposium, IUGG Tsunami Commission, Chania, Crete Island, Greece, June 27–29, 2005, 330 pp.

Papadopoulos, G.A., 1993a. Seismic faulting and nonseismic tsunami generation in Greece. Proceedings of Internaternational Tsunami Symposium, Kyoto University, Wakayama, August 23–27, pp. 115–122.

Papadopoulos, G.A., 1993b. On some exceptional seismic (?) sea-waves in the Greek Archipelago. Sci. Tsunami Hazard 11, 25–34.

Papadopoulos, G.A., 1998. A reconstruction of the 373 B.C. large earthquake in the western Corinthos Gulf. Proceedings of Second International Confererence on the Ancient Eliki, Aeghio, December 1–3, Athens, pp. 479–494.

Papadopoulos, G.A., 2001. Tsunamis in the East Mediterranean: a catalogue for the area of Greece and adjacent seas. Proceedings of Joint IOC-IUGG Internattional. Workshop on Tsunami Risk Assessment Beyond 2000. Theory, Practice and Plans, Moscow, June 14–16, pp. 34–42.

Papadopoulos, G.A., 2003a. Quantification of tsunamis: a review. Advanced Research Workshop Underwater Ground Failures on Tsunami Generation, Modeling, Risk & Mitigation, Kluwer, vol. 229–236, pp. 285–289.

Papadopoulos, G.A., 2003b. Tsunami hazard in the Eastern Mediterranean: strong earthquakes and tsunamis in the Corinth Gulf, Central Greece. Nat. Hazards 29, 437–464.

Papadopoulos, G.A., 2003c. A tsunami warning system in the SW Aegean Sea, Greece. In: Zschau, J., Küppers, A.N. (Eds.), Early Warning Systems for Natural Disaster Reduction. Springer-Verlag, USA, pp. 549–552.

Papadopoulos, G.A., 2009. Tsunamis, Chapter 17. In: Woodward, J. (Ed.), Physical Geography of the Mediterranean. Oxford University Press, Oxford, pp. 493–512.

Papadopoulos, G.A., 2011. A seismic history of Crete: earthquakes and tsunamis, 2000 B.C. – A.D. 2010, Ocelotos Publishers, Athens, 415 pp.

Papadopoulos, G.A., 2014. Rodos – Oi Seismoi kai ta Tsunami apo tin Archaiotita mexri simera. Ocelotos Publ., Athens, ISBN 978-960-564-200-6, 222 pp. (in Greek).

Papadopoulos, G.A., Baskoutas, I., Fokaefs, A., 2014b. Historical seismicity of the Kyparissiakos Gulf, western Peloponnese, Greece. Boll. Geof. Teor. Appl. 55, 389–404.

Papadopoulos, G.A., Caputo, R., McAdoo, B., Pavlides, S., Karastathis, V., Fokaefs, A., Orfanogiannaki, K., Valkaniotis, S., 2006. The large tsunami of 26 December 2004: field observations and eyewitnesses accounts from Sri Lanka, Maldives Is. and Thailand. Earth Planet. Space 58, 233–241.

Papadopoulos, G.A., Chalkis, B., 1984. Tsunamis observed in Greece and the surrounding area from antiquity up to the present times. Mar. Geol. 56, 309–317.

Papadopoulos, G.A., Daskalaki, E., Fokaefs, A., 2007b. Tsunamis generated by coastal and submarine landslides in the Mediterranean Sea: submarine mass movements and their consequences. In: Lykousis, V., et al. (Eds.), Advances in Natural and Technologiacal Hazards, vol. 27, pp. 415–422.
Papadopoulos, G.A., Daskalaki, E., Fokaefs, A., Giraleas, N., 2007a. Tsunami hazard in the Eastern Mediterranean: strong earthquakes and tsunamis in the East Hellenic Arc and Trench System. Nat. Hazards Earth Syst. Sci. 7, 57–64.
Papadopoulos, G.A., Daskalaki, E., Fokaefs, A., Giraleas, N., 2010. Tsunami hazard in the Eastern Mediterranean sea: strong earthquakes and tsunamis in the west Hellenic arc and trench system. J. Earthquakes Tsunamis 4, 145–179.
Papadopoulos, G.A., Dermentzopoulos, Th., 1998. A tsunami risk management pilot study in Heraklion, Crete Isl. Greece. Nat. Hazards 18, 91–118.
Papadopoulos, G.A., Diakogianni, G., Fokaefs, A., Ranguelov, B., 2011. Tsunami hazard in the Black Sea and the Azov Sea: a new tsunami catalogue. Nat. Hazards Earth Syst. Sci. 11, 1–19.
Papadopoulos, G.A., Fokaefs, A., 2005. Strong tsunamis in the Mediterranean Sea: a re-evaluation, ISET. J. Earthquake Technol. 42, 159–170.
Papadopoulos, G.A., Fokaefs, A., 2013. Near-field tsunami early warning and emergency planning in the Mediterranean Sea. Res. Geophys. 3 (1), 4.
Papadopoulos, G.A., Ganas, A., Plessa, A., 2002. The Skyros Earthquake (Mw6.5) of 26 July 2001 and Precursory Seismicity Patterns in the North Aegean Sea. Bull. Seismol. Soc. Am. 92, 1141–1145.
Papadopoulos, G.A., Gràcia, E., Urgeles, R., Sallares, V., De Martini, P.M., Pantosti, D., González, M., Yalciner, A.C., Mascle, J., Sakellariou, D., Salamon, A., Tinti, S., Karastathis, V., Fokaefs, A., Camerlenghi, A., Novikova, T., Papageorgiou, A., 2014a. Historical and pre-historical tsunamis in the Mediterranean and its connected seas: geological signatures, generation mechanisms and coastal impacts. Mar. Geol. 354, 81–109.
Papadopoulos, G.A., Imamura, F., 2001. A proposal for a new tsunami intensity scale. Proceedings of International Tsunami Symposium 2001, Seattle, August 7–10, 2001, pp. 569–577.
Papadopoulos, G.A., Imamura, F., Minoura, K., Takahashi, T., Karakatsanis, S., Fokaefs, A., Orfanogiannaki, K., Daskalaki, E., Diakogianni, G., 2005a. The 9 July 1956 large tsunami in the South Aegean Sea: compilation of a data basis and re-evaluation. In: Papadopoulos, G.A., Satake, K. (Eds.), Proceedings of 22nd International IUGG Tsunami Symposium, June 27–29, Chania, Crete, pp. 173–180.
Papadopoulos, G.A., Imamura, F., Minoura, K., Takahasi, T., Kuran, U., Yalçiner, A., 2005b. Strong earthquakes and tsunamis in the East Hellenic Arc and Trench System. Geophys. Res. Abstr. 6, EGU2005-03212.
Papadopoulos, G.A., Kijko, A., 1991. Maximum likelihood estimation of earthquake hazard parameters in the Aegean area from mixed data. Tectonophysics 185, 277–294.
Papadopoulos, G.A., Kortekaas, S., 2003. Characterisics of landslide gererated tsunamis from observational data. In: Locat, L., Mienert, J. (Eds.), Submarine Mass Movements and Their Consequences. Kluwer, Dordrecht, pp. 367–374.
Papadopoulos, G.A., Lobkovsky, L.I., Mazova, R.Kh., Garagash, I.A., Karastathis, V., Kataeva, L.Yu., Kaz'min, V.G., 2007c. Numerical modeling of sediment mass sliding and tsunami generation: the case of February 7, 1963, in Corinth Gulf, Greece. Mar. Geodesy 30, 315–331.
Papadopoulos, G.A., Minoura, K., Imamura, F., Kuran, U., Yalçiner, A., Fokaefs, A., Takahashi, T., 2012a. Strong earthquakes and tsunamis in the East Hellenic Arc. Res. Geophys. 2 (e12), 90–99.
Papadopoulos, G.A., Novikova, T., Diakogianni, G., 2012b. The large tsunami of 1956 in the South Aegean Sea, Greece, revisited. Presented at the "The Deep-Sea & Sub-Seafloor Frontier Conference, Barcelona, March 11–14, 2012."

Papadopoulos, G.A., Papageorgiou, A., 2014. Large earthquakes and tsunamis in the Mediterranean region and its connected seas. In: Ismail-Zadeh, A., et al. (Ed.), Extreme Natural Hazards, Disaster Risks ans Societal Implications. Cambridge University Press, Cambridge, pp. 252–266.

Papadopoulos, G.A., Daskalaki, E., Fokaefs, A., Novikova, T., 2014c. Tsunamigenic potential of local and distant tsunami sources threatening SW Peloponnese. Boll. Geof. Teor. Appl. 55, 469–484.

Papadopoulos, G.A., Pavlides, S., 1992. The 1956 large shock in the South Aegean: macroseismic field, faulting and neotectonics of the Amorgos Island. Earth Planet. Sci. Lett. 113, 383–396.

Papadopoulos, G.A., Vassilopoulou, A., Plessa, A., 2000. A new catalogue of historical earthquakes in the Corinth Rift, Central Greece: 480 B.C.–A.D. 1910. In: Papadopoulos, G.A. (Ed.), Historical earthquakes and tsunamis in the Corinth Rift, Central Greece, Institute of Geodynamics, National Observatory Athens, Publication No. 12, pp. 9–119.

Papageorgiou, A., Tsimi, Ch., Orfanogiannaki, K., Papadopoulos, G.A., Sachpazi, M., Lavigne, F., Grancher, D., 2015. Tsunami Questionnaire Survey in Heraklion Test Site, Crete Island, Greece. Geophys. Res. Abstr., 17, EGU2015-10784, EGU General Assembly 2015.

Papaioannou, I., Papadopoulos, G.A., Pavlides, S., 2004. The earthquake of 426 BC in N. Evoikos Gulf revisited: amalgamation of two different strong earthquake events? Bull. Geol. Soc. Greece 36, 1477–1481.

Papathoma, M., 2003. Tsunami vulnerability assessment using a Geographical Information System with special reference to Greece. PhD Thesis, Coventry University, March 2003, 289 pp.

Papathoma, M., Dominey-Howes, D., 2003. Tsunami vulnerability assessment and its implications for coastal hazard analysis and disaster management planning, Gulf of Corinth, Greece. Nat. Hazards Earth Syst. Sci. 3, 733–747.

Papathoma, M., Dominey-Howes, D., Zong, Y., Smith, D., 2003. Assessing tsunami vulnerability, an example from Herakleio, Crete. Nat. Hazards Earth Syst. Sci. 3, 1–13.

Papazachos, B.C., Dimitriu, P.P., 1991. Tsunamis in and near Greece and their relation to the earthquake focal mechanisms. Nat. Hazards 4, 161–170.

Papazachos, B.C., Koutitas, C., Hatzidimitriou, P., Karacostas, B., 1956. Papaioannou, Ch. 1985. Source and short-distance propagation of the July 9, 1956 southern Aegean tsunami. Mar. Geol. 65, 343–351.

Papazachos, B.C., Koutitas, C., Hatzidimitriou, P., Karacostas, B., Papaioannou, Ch., 1985. Source and short-distance propagation of the July 9, 1956 southern Aegean tsunami. Mar. Geol. 65, 343–351.

Papazachos, B.C., Koutitas, Ch., Hatzidimitriou, P.M., Karakostas, B.M., Papaioannou, Ch.A., 1986. Tsunami hazard in Greece and the surrounding area. Annales Geophysicae 4, 79–90.

Papoulia, J., Makris, J., Mascle, J., Slejko, D., Yalçiner, A., 2014. The EU SEAHELLARC project: aims and main results. Boll. Geof. Teor. Appl. 55, 241–248.

Pararas-Carayannis, G., 1992. The tsunami generated from the eruption of the volcano of Santorin in the Bronge Age. Nat. Hazards 5, 115–123.

Pareschi, M., Favalli, M., Boshi, E., 2006a. Impact of the Minoan tsunami of Santorini: simulated scenarios in the eastern Mediterranean. Geophys. Res. Lett. 33, 1–6.

Pareschi, M., Favalli, M., Boshi, E., 2006c. Impact of the Minoan tsunami of Santorini: simulated scenarios in the eastern Mediterranean. Geophys. Res. Lett. 33, 1–6.

Pareschi, M.T., Boschi, E., Favalli, M., 2006a. Lost tsunami. Geophys. Res. Lett. 33, L22608.

Pareschi, M.T., Boschi, E., Favalli, M., 2007. Holocene tsunamis from Mount Etna and the fate of Israeli Neolithic communities. Geophys. Res. Lett. 34, L16317.

Pareschi, M.T., Boschi, E., Favalli, M., 2008. Reply to comments by E. Galili et al. on "Holocene tsunami's from Mount Etna and the fate of Israeli Neolithic communities". Geophys. Res. Lett. 35, L08312.

Pareschi, M.T., Boschi, E., Mazzarini, F., Favalli, M., 2006b. Large submarine landslides offshore Mt. Etna. Geophys. Res. Lett. 33, L13302.

Paris, R., Fournier, J., Poizot, E., Etienne, S., Mortin, J., Lavigne, F., Wassmer, P., 2010. Boulder and fine sediment transport and deposition by the 2004 tsunami in Lhok Nga (western Banda Aceh, Sumatra, Indonesia): a coupled offshore-onshore model. Mar. Geol. 268, 43–54.

Paris, R., Wassmer, P., Sartohadi, J., Lavigne, F., Barthomeuf, B., Desgages, E., Grancher, D., Baumert, Ph., Vautier, F., Brunstein, D., Gomez, Ch., 2009. Tsunamis as geomorphic crisis: lessons from the December 26, 2004 tsunami in Lhok Nga, west Banda Aceh (Sumatra, Indonesia). Geomorphology 104, 59–72.

Pasarić, M., Brizuela, B., Graziani, l., Maramai, A., Orlić, M., 2012. Historical tsunamis in the Adriatic Sea. Nat. Hazards 61, 281–316.

Patacca, E., Scandone, P., 2004. The 1627 Gargano earthquake (Southern Italy): identification and characterization of the causative fault. J. Seismol. 8, 259–273.

Paulatto, M., Pinat, T., Romanelli, F., 2007. Tsunami hazard senarios in the Adriatic Sea domain. Nat. Hazards Earth Syst. Sci. 7, 309–325.

Peiris, N., Pomonis, A., 2006. Vulnerability functions for tsunami loss estimation. First European conference on Earthquake Engineering and Seismology (a joint event of the 13thECEE and 30th General Assembly of the ESC), Geneva, Switzerland, Paper number 1121.

Pelinovsky, E., 1999. Preliminary estimates of tsunami danger for the northern part of the Black Sea. Phys. Chem. Earth 24, 175–178.

Periáñez, R., Abril, J.M., 2014. Modelling tsunamis in the Eastern Mediterranean Sea. Application to the Minoan Santorini tsunami sequence as a potential scenario for the biblical exodus. J. Marine Syst. 139, 91–102.

Perissoratis, C., Papadopoulos, G.A., 1999. Sediment instability and slumping in the southern Aegean Sea and the case history of the 1956 tsunami. Mar. Geol. 161, 287–305.

Piatanesi, A., Tinti, S., Bortolucci, E., 1999. Finite-Element Simulations of the 28 December 1908 Messina Straitis (Southern Italy) Tsunami. Phys. Chem. Earth 2 (24), 144–150.

Pino, N., Giardini, D., Boschi, E., 2000. The December 28, 1908, Messina Straits, southern Italy, earthquake: waveform modeling of regional seismograms. J. Geophys. Res. 105, 25473–25492.

Pino, N.A., Piatanesi, A., Valensise, G., Boschi, E., 2009. The 28 December 1908 Messina Straits earthquake (Mw 7.1): a great earthquake throughout a century of seismology. Seism. Res. Lett. 80 (2), 243–259.

Pino, N.A., Piatanesi, A., Valensise, G., Boschi, E., 2010. Reply to "Comment on 'The 28 December 1908 Messina Straits Earthquake (Mw 7.1): a great earthquake throughout a century of seismology,' by N. A. Pino, A. Piatanesi, G. Valensise, and E. Boschi" by Amoruso, A., Crescentini, L., Scarpa, R. Seism. Res. Lett., 81, 229–231.

Pirazzoli, P.A., 2010. Scientific report on the COST-STSM-ES0701-6735. http://www.cost-es0701.geoenvi.org/attachments/article/202/Pirazzoli_STSM_2010 Report.pdf.

Pirazzoli, P.A., Ausseil-Badie, J., Giresse, P., Hadjidaki, E., Arnold, M., 1992. Historical environmental changes at Phalasarna harbour, West Crete. Geoarchaeology 7, 371–392.

Piscini, A., Maramai, A., Tinti, S., 1998. Pilot local tsunami warning system in Augusta, eastern Sicily, Italy. Proc. Internat. Conf. on Tsunamis, Paris, May 26–28, 1998, pp. 137–150.

Platania, G., 1909. I fenomeni marittimi che accompagnarono il terremoto di Messina del 28 dicembre 1908. Rivista Geografica Italiana 16, 154–161.

Polonia, A., Bonatti, E., Carmelenghi, A., Lucchi, R.G., Panieri, G., Gasperini, L., 2013. Mediterranean megaturbidite triggered by the AD 365 Crete earthquake and tsunami. Scientific Reports 3, 1285.

Polonia, A., Panieri, G., Gasperini, L., Gasparotto, G., Bellucci, L.G., Torelli, L., 2012. Turbidite paleoseismology in the Calabrian Arc Subduction Complex (Ionian Sea). Geochem. Geophys. Geosyst. 14 (1).

Pondrelli, S., Ekström, G., Morelli, A., Primerano, S., 1998. Study of source geometry for tsunamigenic events of the Euro-Mediterranean area. Proc. Internat. Conf. on Tsunamis, Paris, May 26–28 1998, 297–307.

Preuss, J., 1991. Urban planning for tsunami hazards Grays harbor, Washington and Lima, Peru. In: Brennan, A.M., Lander, J.F. (Eds.), 2nd UJNR Tsunami Workshop, November 5–6, 1990, Honolulu, Hawaii.

Pryor, J.H., 2006. Logistics of warfare in the Age of the Crusades. Proceedings of a Workshop held at the Centre for Medieval Studies, University of Sydney, September 30 to October 4, 2002. Ashgate Publishing. p. 115. ISBN 978-0-7546-5197-0.

Raicich, F., Orlić, M., Vilibić, I., Malačič, V., 1999. A case study of the Adriatic seiches (December 1997). Nuovo Cimento C22, 715–726.

Rangin, C., Le Pichon, X., Demirbag, E., Imren, C., 2004. Strain localization in the Sea of Marmara: propagation of the North Anatolian Fault in a now inactive pull- apart. Tectonics 23, TC2014.

Ranguelov, B., Jelev, G., Novikova, T., et al., 2010. Atlas of the tsunami-risk-susceptible areas along the northern Bulgarian Black Sea coast- Balchik site. SCHEMA EU FP6 Project., Space Research Institute, Bulgarian Academy of Sciences & National Observatory of Athens, Greece, 23 pp.

Ranguelov, B., Mircheva, E., Lazarenko, I., Encheva, R., 2008a. The archaeological site- possible evidence about multihazard ancient events. Proceedings of Conference of Geo-archaeology and Archaelomineralogy, 2008 Sofia, Publishing House "St. Ivan Rilski," Sofia, pp. 347–352.

Ranguelov, B., Tinti, S., Pagnoni, G., Tonini, R., Zaniboni, F., Armigliato, A., 2008b. The nonseismic tsunami observed in the Bulgaria Sea on 7 May 2007 – Was it due to a sub-marine landslide? Geophys. Res. Lett. 35, L18613.

Reese, L.C., Matlock, H., Asce, M., 1968. Structural Damage from Tsunami at Hilo, Hawaii. J. Hydraul. Div. 94 (4), 961–982, July/August 1968.

Reese, S., Cousins, W.J., Power, W.L., Palmer, N.G., Tejakusuma, I.G., Nugrahadi, S., 2007. Tsunami vulnerability of buildings and people in South Java? Field observations after the July 2006 Java tsunami. Nat. Hazards Earth Syst. Sci. 7, 573–589.

Reicherter, K., Becker-Heidmann, P., 2009. Tsunami deposits in the western Mediterranean: remains of the 1522 Almeria earthquake? In: Reicherter, K., Michetti, A.M., Silva, P.G. (Eds.), Palaeoseismology: Historical and Prehistorical Records of Earthquake Ground Effects for Seismic Hazard Assessment, vol. 316, Geological Society London Special Publication, London, pp. 217–235.

Reicherter, K., Papanikolaou, I., Roger, J., Mathes-Schmidt, M., Papanikolaou, D., Rössler, S., Grützner, Ch., Stamatis, G., 2010. Holocene tsunamigenic sediments and tsunami modeling in the Thermaikos Gulf area (northern Greece). Zeitschrift für Geomorphologie 54 (3), 99–126.

Renou, C., Lesne, O., Mangin, A., Rouffi, F., Atillah, A., El Hadani, D., Moudni, H., 2011. Tsunami hazard assessment in the coastal area of Rabat and Salé, Morocco. Nat. Hazards Earth Syst. Sci. 11, 2181–2191.

Ribeiro, J., Silva, A., Leitão, P., 2011. High resolution tsunami modelling for the evaluation of potential risk areas in Setúbal (Portugal). Nat. Hazards Earth Syst. Sci. 11, 2371–2380.

Richter, C.F., 1935. An instrumental earthquake scale. Bull. Seism. Soc. Am. 25, 1–32.

Ripepe, M., Marchetti, E., 2004. Seismic, acoustic, and thermal network monitors the 2003 eruption of Stromboli volcano. EOS Transactions, AGU 85 (35), 330–332.

Roger, J., Hébert, H., 2008. The 1856 Djidjelli (Algeria) earthquake and tsunami: source parameters and implications for tsunami hazard in the Balearic Islands. Nat. Hazards Earth Syst. Sci. 8, 721–731.

Rolland, L.M., Lognonné, Astafyeva, E., et al., 2011. The resonant response of the ionosphere imaged after the 2011 off the Pacific coast of Tohoku earthquake. Earth Planets Space 63, 853–857.

Römer, H., Willroth, P., Kaiser, G., Vafeidis, A.T., Ludwig, R., Sterr, H., Diez, J.R., 2012. Potential of remote sensing techniques for tsunami hazard and vulnerability analysis- a case study from Phang- Nga province. Thailand. Nat. Hazards Earth Syst. Sci. 12, 2103–2126.

Romstad, B., Harbitz, C.B., Domaas, U., 2009. A GIS method for assessment of rock slide tsunami hazard in all Norwegian lakes and reservoirs. Nat. Hazards Earth Syst. Sci. 9, 353–364.

Rosser, J., 1985. Excavations at Saranda Kolones, Paphos, Cyprus, 1981-1983 (in French) (abstr.). Dumbarton Oaks Papers 39, 81–97.

Rothwell, R.G., et al., 1998. Low-sea-level emplacement of a very large Late Pleistocene "megaturbidite" in the western Mediterranean Sea. Nature 392, 377–380.

Rudloff, A., Lauterjung, J., Münch, U., Tinti, S., 2009. Preface "The GITEWS Project (German-Indonesian Tsunami Early Warning System)". Nat. Hazards Earth Syst. Sci. 9, 1381–1382.

Ruiz, F., Abad, M., Cáceres, L.M., Vidal, J.R., Carretero, M.I., Pozo, M., González-Regalado, M.L., 2010. Ostracods as tsunami tracers in Holocene sequences. Quat. Res. 73, 130–135.

Ruiz, F., Rodríguez-Vidal, J., Abad, M., Cáceres, L.M., Carretero, M.I., Pozo, M., Rodríguez-Lanes, J.M., Gómez-Toscano, F., Izquierdo, T., Font, E. and Toscanoa, A., 2013. Sedimentological and geomorphological imprints of Holocene tsunamis in southwestern Spain: an approach to establish the recurrence period Dumbarton Oaks Papers, 39, pp. 81–97.

Ryan, W.B.F., 2008. Decoding the Mediterranean salinity crisis. Sedimentology 56, 95–136.

Sahal, A., Lemahieu, A., 2011. The 1979 Nice airport tsunami: mapping of the flood in Antibes. Nat. Hazards 56, 833–840.

Sahal, A., Leone, F., Péroche, M., 2013. Complementary methods to plan pedestrian evacuation of the French Riviera's beaches in case of tsunami threat: graph- and multi-agent-based modelling. Nat. Hazards Earth Syst. Sci. 13, 1735–1743.

Sahal, A., Roger, J., Allgeyer, S., et al., 2009. The tsunami triggered by the 21 May 2003 Boumerdès-Zemmouri (Algeria) earthquake: field investigations on the French Mediterranean coast and tsunami modelling. Nat. Hazards Earth Syst. Sci. 9, 1823–1834.

Sakellariou, D., Rousakis, G., Nomikou, P., Croff Bell, K., Carey, S., Sigurdsson, H., 2012. Tsunami Triggering Mechanisms Associated with the 17th century BC Minoan Eruption of Thera Volcano, Greece. Proceedings of 22nd International Offshore and Polar Engineering Conference, Tsunami and Safety Symposium, Rhodes, Greece, June 17–22, 2012, pp. 61–67.

Salamon, A., 2004. Seismically induced ground effects of the February 11, 2004, ML = 5.2, northeastern Dead Sea earthquake. Geol. Survey of Israel, Report GSI/30/2004, December 2004, 20 pp.

Salamon, A., 2011. Potential tsunamigenic sources in the eastern Mediterranean and a decision matrix for a tsunami early warning system. In: Briand, F. (Ed.), Marine Geo-Hazards in the Mediterranean, Workshop Monographs, Rapports et Procés-Verbaux des Réunions de la Commission Internationale pour l'Exploration Scientifique de la Mer Méditeranée (CIESM), Monaco 42, pp. 35–42.

Salamon, A., Rockwell, T., Guidoboni, E., Comastri, A., 2009. A critical evaluation of tsunami records reported for the Levant coast from the second millennium BCE to the present, Israel. Israel J. Earth Sci. 58 (3–4), 327–354.

Salamon, A., Rockwell, T., Ward, S.N., Guidoboni, E., Comastri, A., 2007. Tsunami hazard evaluation of the eastern Mediterranean: historical analysis and selected modeling. Bull. Seismol. Soc. Am. 97 (3), 705–724.

Sasaki S, 2012. The great east Japan tsunami and tsunami warning systems: policy perspectives- What happened at that time? Report from affected areas. In: Japan–IOC/UNESCO/ UNU Symposium "The Great East Japan Tsunami on 11 March 2011 and Tsunami

Warning Systems: Policy Perspectives", 2012 Feb 16-17, UN University, Tokyo, Japan. Available from: http://ioc-tsunami.org/index.php?option= com_oe&task=view Document Record & doc ID=8606.
Sasorova, E., Korovin, M.E., Morozov, V.E., Savochkin, P.V., 2008. On the problem of local tsunamis and possibilities of their warning. Oceanology 48, 634–645.
Satake, K., Fuji, Y., Harada, T., 2011. Tsunami source model of the 2011 Tohoku earthquake inferred from tsunami waveforms and coastal inundation heights. Abstract NH14A-01 presented at 2011 Fall Meeting AGU, San Francisco, Calif., December 5–9.
Satake, K., Okal, E., Borrero, J.C., 2007. Tsunami and its hazards in the Indian and Pacific Oceans. Pure Appl. Geophys. 164, 249–631.
Scheffers, A., Kelletat, D., 2005. Tsunami relics on the coastal landscape west of Lisbon, Portugal. Sci. Tsunami Hazards 1 (23), 3.
Scheffers, A., Kelletat, D., Vött, A., May, S.M., Scheffers, S., 2008. Late Holocene tsunami traces on the western and southern coastlines of the Peloponnesus (Greece). Earth Planet. Sci. Lett. 269, 271–279.
Scheffers, A., Scheffers, S., 2007. Tsunami deposits on the coastline of west Crete (Greece). Earth Planet. Sci. Lett. 259, 613–624.
Schettino, A., Turco, E., 2009. Breakup of Pangea and plate kinematics of the central Atlantic and Atlas regions. Geophys. J. Int. 178, 1078–1097.
Schindelé, F. 1998. Tsunami warning in near-field for the two large 1996 Peru earthquakes. Proceedings of International Conference on Tsunamis, 1998 May 26–28, Paris, pp. 173–184.
Schindelé, F., Loevenbruck, A., Hébert, H., 2008. Strategy to design the sea-level monitoring networks for small tsunamigenic oceanic basins: the Western Mediterranean case. Nat. Hazards Earth Syst. Sci. 8, 1019–1027.
Schindelé, F., Roch, J., Rivera, L., 2015. Preliminary results of very fast computation of Moment Magnitude and focal mechanism in the context of tsunami warning. Geophys. Res. Abstr., 17, EGU2015-2926-1, EGU General Assembly 2015.
Schmidt, J.F., 1879. Studien über Erdbeben. Carl Scholtze, Leipzig, 360 pp.
Schuiling, R.D., Cathcart, R.B., Badescu, V., Isvoranu, D., Pelinovsky, E., 2007. Asteroid impact in the Black Sea. Death by drowning or asphyxiation? Nat. Hazards 40, 327–338.
Šepić, J., Vilibić, I., 2011. The development and implementation of a real-time meteotsunami warning network for the Adriatic Sea. Nat. Hazards Earth Syst. Sci. 11, 83–91.
Serri, G., 1997. Neogene-Quaternary magmatic activity and its geodynamic implications in the central Mediterranean region. Ann. Geofis. XL (3), 681–703.
Shaw, B., Ambraseys, N.N., England, P.C., Floyd, M.A., Gorman, G.J., Higham, T.F.G., Jackson, J.A., Nocquet, J.-M., Pain, C.C., Piggott, M.D., 2008. Eastern Mediterranean tectonics and tsunami hazard inferred from the AD 365 earthquake. Nat. Geosci. 1, 268–276.
Shuto, N., 1987. The effectiveness and limit of tsunami control forests. Coast. Eng. Japan 30, 143–153.
Shuto, N., 1991. Historical changes in characteristics of tsunami disasters. Proceedings of International Symposium on Natural Disaster Reduction and Civil Engineering, pp. 77–86.
Shuto, N., 1993. Tsunami intensity and disasters. In: Tinti, S. (Ed.), Tsunamis in the World. Kluwer, Dordrecht, pp. 197–216.
Sieberg, A., 1923. Geologische, physicalische und angewandte Erdbebenkunde. Verlag von G. Fischer, Jena.
Sieberg, A., 1927. Geologische, physikalische und angewandte Erdbebenkunde. Verlag von Gustav Fischer, Jena, 374 pp.
Sieh, K., Natawidjaja, D.H., Meltzner, A.J., et al., 2008. Earthquake supercycles inferred from sea-level changes recorded in the corals of west Sumatra. Science 322 (5908), 1674–1678.
Sigurdsson, H., Carey, S., Devine, J.D., 1990. Assessment of mass, dynamics and environmental effects of the Minoan eruption of Santorini volcano. In: Hardy, D.A., Keller, J.,

Galanopoulos, V.P., Flemming, N.C., Druitt, T.H. (Eds.), Thera and the Aegean World III, 2, The Thera Foundation, London, pp. 100–112.
Sim, S.Y., Huang, Z., Switzer, A.D., 2013. An experimental study on tsunami inundation over complex coastal topography. Theoretical Appl. Mech. Lett. 3 (3), 9–03200.
Singhal, A., Kiremidjian, A.S., 1996. Method for probabilistic evaluation of seismic structural damage. J. Struct. Eng. 122, 1459–1467.
Smedile, A., De Martini, P.M., Pantosti, D., Bellucci, L., Del Carlo, P., Gasperini, L., Pirrotta, C., Polonia, A., Boschi, E., 2011. Possible tsunamis signatures from an integrated study in the Augusta Bay offshore (Eastern Sicily-Italy). Mar. Geol. 281, 1–13.
Smith, K., 1992. Environmental Hazards – Assessing Risk & Reducing Disaster. Routledge, United Kingdom, 324 pp.
Società Fotografica Italiana, 2002. Messina e Reggio: prima e dopo il Terremoto del 28 1908. Nuove "Edizioni Barbaro," Delianuova, 437 pp.
Solares, J.M.M., Arroyo, A.L., 2004. The great historical 1755 earthquake. Effects and damage in Spain. J. Seismol. 8, 275.
Soloviev, S.L., 1970. Recurrence of tsunamis in the Pacific. In: Adams, W.M. (Ed.), Tsunamis in the Pacific Ocean. East-West Center Press, Honolulu, pp. 149–163.
Soloviev, S.L., 1990. Tsunamigenic zones in the Mediterranean Sea. Nat. Hazards 3, 183–202.
Soloviev, S.L., Solovieva, O.N., Go, C.N., Kim, K.S., Shchetnikov, N.A., 2000. Tsunamis in the Mediterranean Sea 2000 B.C.–2000 A.D. Advances in Natural and Technological Hazards Research 13. Kluwer Academic Publishers, London.
Somerville, P., 2012. Aftermath of the 2011 Tohoku, Japan earthquake. Risk Frontiers Newslett. 11 (4), 1–4.
Soren, D., 1988. The day the world ended at Kourion, Reconstructing an ancient earthquake. Natl Geogr. 174, 30–53.
Sørensen, M.B., Spada, M., Babeyko, A., Wiemer, S., Grünthal, G., 2012. Probabilistic tsunami hazard in the Mediterranean Sea. J. Geophys. Res. 117, B01305.
Stefatos, A., Charalambakis, M., Papatheodorou, G., Ferentinos, G., 2006. Tsunamigenic sources in an active European half-graben (Gulf of Corinth, Central Greece). Mar. Geol. 232, 35–47.
Stein, S., Okal, E., 2005. Speed and size of the Sumatra earthquake. Nature 434 (7033), 581–582.
Stich, D., et al., 2007. Source analysis of the February 12th, Mw6. 0 Horseshoe earthquake: implications for the 1755 Lisbon earthquake. Geophys. Res. Lett. 34, L12308.
Stich, D., Mancilla, F., Morales, J., 2005. Crust mantle coupling in the Gulf of Cadiz (SW Iberia). Geophys. Res. Lett. 32, L13306.
Suenaga M., 2012. Miracle in Kamaishi really? Disaster education in school and community. How to do it. In: Japan–IOC/UNESCO/UNU Symposium "The Great East Japan Tsunami on 11 March 2011 and Tsunami Warning Systems: Policy Perspectives," Feb 16–17, 2012. UN University, Tokyo, Japan. Available from: http://ioc-tsunami.org/index.php?option=com_oe&task=viewDocumentRec ord&doc ID=8614.
Sugioka, H., Hamano, Y., Baba, K., Kasaya, T., Tada, N., Suetsugu, D., 2014. Tsunami: ocean dynamo generator. Scientific Reports, 4, 3596, DOI: 10.1038/srep03596.
Suppasri, A., Mas, E., Charvet, I., Gunasekera, R., Imai, K., Fukutani, Y., Abe, Y., Imamura, F., 2013. Building damage characteristics based on surveyed data and fragility curves of the 2011 Great East Japan tsunami. Nat. Hazards 66, 319–341.
Suzuki, W., Aoi, S., Sekiguchi, H., Kunugi, T., 2011. Rupture process of the 2011 Tohoku-Oki mega-thrust earthquake (M9.0) inverted from strong-motion data. Geophys. Res. Lett. 38, L005G16.
Takahashi, T., 2006. Concept of tsunami detecting system using remote sensing. Proceedings of Second International Workshop on Coastal Disaster Prevention, January 18–19, 2006, Tokyo, Japan.

Tanaka, G., Naruse, H., Yamashita, S., Arai, K., 2012. Ostracodes reveal the sea-bed origin of tsunami deposits. Geophys. Res. Lett. 39, L05406.

Tanioka, Y., Satake, K., 1996. Fault parameters of the 1896 Sanriku tsunami earthquake estimated from tsunami numerical modeling. Geophys. Res. Lett. 23, 1549–1552.

Tarbotton, C., Dall'Osso, F., Dominey-Howes, D., Goff, J., 2015. The use of empirical vulnerability functions to assess the response of buildings to tsunami impact: comparative review and summary of best practice. Earth Sci. Rev. 142, 120–134.

Taubenböck H., Post J., Kiefl R., Roth A., Ismail F.A., Strunz, G., Dech S., 2008. Risk and vulnerability assessment to tsunami hazard using very high resolution satellite data. In: Carsten Jürgens (Ed.), Remote Sensing – New challenges of High Resolution, Bochum 2008, pp. 77–86.

Terrinha, P., et al., 2003. Tsunamigenic-seismogenic structures, neotectonics, sedimentary processes and slope instability on the southwest Portuguese Margin. Mar. Geol. 195 (1–4), 55–73.

Terrinha, P., et al., 2009. Morphotectonics and strain partitioning at the Iberia–Africa plate boundary from multibeam and seismic reflection data. Mar. Geol. 267, 156–174.

Theilen-Willige, B., 2008. Tsunami hazard assessment in the northern Aegean Sea. Sci. Tsunami Hazards 27, 1–16.

Thiebot, E., Gutscher, M.A., 2006. The Gibraltar Arc seismogenic zone (part 1): constraints on a shallow east dipping fault plane source for the 1755 Lisbon earthquake provided by seismic data, gravity and thermal modeling. Tectonophysics 426, 135–152.

Thommeret, Y., Thommeret, J., Laborel, J., Montaggioni, L.F., Pirazzoli, P.A., 1981. Late Holocene shoreline changes and seismo-tectonic displacements in western Crete (Greece). Zeitschrift für Geomorphologie. Supplement 40, 127–149.

Tiberti, M., Lorito, S., Basili, R., Kastelic, V., Piatanesi, A., Valensise, G., 2008. Scenarios of earthquake-generated tsunamis for the Italian coast of the Adriatic Sea. Pure Appl. Geophys. 165, 2117–2142.

Tinti, S. (Ed.), 2005. ISET. J. Earthquake Technol. 42(4), 79–236, Special Issue: Tsunami Seismic Hazard.

Tinti, S., 1990. Contributions expected from marine geodesy to the study of tsunamis in the Mediterranean Sea. Mar. Geodesy 14, 243–254.

Tinti, S., 1991. Assessment of tsunami hazard in the Italian Seas. Nat. Hazards 4, 267–283.

Tinti, S., 1993. Evaluation of tsunami hazard in Calabria and eastern Sicily, Italy. In: Tinti, S. (Ed.), Tsunamis in the World. Kluwer, Dordrecht, pp. 141–157.

Tinti, S., Armigliato, A., 2003. The use of scenarios to evaluate tsunami impact in South Italy. Mar. Geol. 199, 221–243.

Tinti, S., Armigliato, A., Manucci, A., Pagnoni, G., Zaniboni, F., Yalçiner, A.C., Altinok, Y., 2006b. The generating mechanisms of the August 17, 1999 Izmit bay (Turkey) tsunami: regional (tectonic) and local (mass instabilities) causes. Mar. Geol. 225, 311–330.

Tinti, S., Armigliato, A., Pagnoni, G., Zaniboni, F., 2005b. Scenarios of giant tsunamis of tectonic origin in the Mediterranean. ISET J. Earthquake Technol. 42/4, 171–188.

Tinti, S., Armigliato, A., Tonini, R., Maramai, A., Graziani, L., 2005a. Assessing the hazard related to tsunamis of tectonic origin: a hybrid statistical-deterministic method applied to southern Italy coasts. ISET J. Earthquake Technol. 42/4, 189–201.

Tinti, S., Armigliato, Bortolucci, E., Platanesi, A., 1999. Identification of the Source Fault of the 1908 Messina Earthquake through Tsunami Modelling. Is it a Possible Task? Phys. Chem. Earth 5 (24), 417–421.

Tinti, S., Canals, M., Pagnoni, G., Zaniboni, F., Iglesias, O., Lastras, G., 2009. The BIG'95 event, Balearic Islands, Western Mediterranean Sea: numerical simulation of the possibly generated tsunami. EGU Gen. Ass., Vienna, April 19–24, 2009, abstr. n. EGU2009-13365.

Tinti, S., Graziani, L., Brizuela, B., Maramai, A., Gallazzi, S., 2012. Applicability of the decision matrix of North Eastern Atlantic, Mediterranean and connected seas tsunami warning system to the Italian tsunamis. Nat. Hazards Earth Syst. Sci. 12, 843–857.

Tinti, S., Guidoboni, E., 1988. Revision of the tsunamis occurred in 1783 in Calabria and Sicily (Italy). Sci. Tsunami Hazards 6, 17–22.

Tinti, S., Manucci, A., Pagnoni, G., Armigliato, A., Zaniboni, F., 2005c. The 30 December 2002 landslide-induced tsunamis in Stromboli: sequence of the events reconstructed from the eyewitness accounts. Nat. Hazards Earth Syst. Sci. 5, 763–775.

Tinti, S., Maramai, A., 1996. Catalogue of tsunamis generated in Italy and in Cote d'Azur, France: a step towards a unified catalogue of tsunamis in Europe. Ann. Geofis. 39, 1253–1299.

Tinti, S., Maramai, A., Armigliato, A., Graziani, L., Manucci, A., Pagnoni, G., Zaniboni, F., 2006a. Observations of physical effects from tsunamis of December 30, 2002 at Stromboli volcano, southern Italy. Bull. Volcanol. 68, 450–461.

Tinti, S., Maramai, A., Favali, P., 1995. The Gargano promontory: an important Italian seismogenic-tsunamigenic area. Mar. Geol. 122 (2), 227–241.

Tinti, S., Maramai, A., Graziani, L., 2004. The new catalogue of Italian tsunamis. Nat. Hazards 33, 439–465.

Tinti, S., Pagnoni, G., Zaniboni, F., 2005d. The landslides and tsunamis of the 30th of December 2002 in Stromboli analysed through numerical simulations. Bull. Volcanol. 68, 462–479.

Tinti, S., Pagnoni, G., Zaniboni, F., Bortolucci, E., 2003. Tsunami generation in Stromboli island and impact on the south-east Tyrrhenian coasts. Nat. Hazards Earth Syst. Sci. 3, 299–309.

Tinti, S., Piatanesi, A., 1996. Numerical simulations of the tsunami induced by the 1627 earthquake affecting Gargano, southern Italy. J. Geodyn. 21, 141–160.

Tinti, S., Tonini, R., 2013. The UBO-TSUFD tsunami inundation model: validation and application to a tsunami case study focused on the city of Catania, Italy. Nat. Hazards Earth Syst. Sci. 13, 1795–1816.

Tinti, S., Zaniboni, F., Armigliato, A., Pagnoni, G., Gallazzi, S., Manucci, A., Brizuela Reyes, B., Bressan, L., Tonini, R., 2007. Tsunamigenic landslides in the western Corinth Gulf: numerical simulations. In: Lykousis, V. et al.,(Ed.), Submarine Mass Movements and Their Consequences. Springer, Vienna, pp. 405–414.

Tonini, R., Armigliato, A., Pagnoni, G., Zaniboni, F., Tinti, S., 2011. Tsunami hazard for the city of Catania, eastern Sicily, Italy, assessed by means of Worst-case Credible Tsunami Scenario Analysis (WCTSA). Nat. Hazards Earth Syst. Sci. 11, 1217–1232.

Triantafyllou, I., Nomikou E., Papadopoulos, G.A., 2015. Historical tsunamis in the Island volcanic complex of Thera (Santorini), Greece. 26th IUGG General Assembly, Prague, June–July 2015. https://www.czechin.org/cm/IUGG/CM.NET. WebUI/CM.NET.WEBUI.scpr/SCPRsessions.aspx?conferenceid=05000000-0000-0000-0000000000000053&sessionID =05000000-0000-0000-0000-000000003642.

Tselentis, A., Gkika, F., Sokos, E., 2006. Tsunami hazards associated with the Perachora fault at eastern Corinth Gulf, Greece. Bull. Seism. Soc. Am. 96, 1649–1661.

Tselentis, G.-A., Stavrakakis, G., Sokos, E., Gkika, F., Serpetzidaki, A., 2010. Tsunami hazard assessment in the Ionian Sea due to potential tsunamogenic sources – results from numerical simulations. Nat. Hazards Earth Syst. Sci. 10, 1021–1030.

Tsimplis, M.N., Blackman, D., 1997. Extreme sea-level distribution and return periods in the Aegean and Ionian Seas. Est. Coast. Shelf Sci. 44, 79–89.

Tsuji, Y., Matsutomi, H., Tanioka, Y., et al., 2005. Distribution of the tsunami heights of the 2004 Sumatra in Banda Aceh measured by the tsunami survey team. http://www.eri.u-tokyo.ac.jp/namegaya/sumatera/surveylog/eindex.htm.

Ulutas, E., Inan, A., Annunziato, A., 2012. Web-based tsunami early warning system: a case study of the 2010 Kepulaunan Mentawai earthquake and tsunami. Nat. Hazards Earth Syst. Sci. 12, 1855–1871.

UNESCO-IOC (Intergovernmental Oceanographic Commission), 2011. Reducing and man-aging the risk of tsunamis. IOC Manuals and Guides (IOC/2011/MG/57Rev.2)), pp 57–74.

UNESCO, 2015. From early warning to early response-tsunami exercise NEAMWave 14-Summary of Results. UNESCO, IOC/BRO/2015/7, 13 pp.
Urgeles, R., Camerlenghi, A., 2013. Submarine landslides of the Mediterranean Sea: trigger mechanisms, dynamics and frequency-magnitude distribution. J. Geophys. Res. Earth Surf. 118, 2600–2618. http://dx.doi.org/10.1002/2013JF002720.
Urgeles, R., et al., 2007. Sediment undulations on the Llobregat prodelta: signs of early slope instability or bottom current activity? J. Geophys. Res. 112 (B5), Art. No. B05102.
USACE, 1990. Wave forces on a wall shoreward of the still-water line. Coastal Engin. Technical Note-III-29. USACE Waterways Experiment Station, Mississippi, USA.
Vacchi, M., Rovere, A., Zouros, N., Firpo, M., 2012. Assessing enigmatic boulder deposits in NE Aegean Sea: importance of historical sources as tool to support hydrodynamic equations. Nat. Hazards Earth Syst. Sci. 12, 1109–1118.
Valencia, N., Gardi, A., Gauraz, A., Leone, F., Guillande, R., 2011. New tsunami damage functions developed in the framework of SCHEMA project: application to European-Mediterranean coasts. Nat. Hazards Earth Syst. Sci. 11, 2835–2846.
Valensise, G., Pantosti, D., 1992. A125 kyr- long geological record of seismic source repeatability: the Messina Straits (southern Italy) and the 1908 earthquake (Ms 7½). Terra Nova 4, 472–483.
Vallée, M., 2007. Rupture properties of the giant Sumatra earthquake imaged by empirical Green's function analysis. Bull. Seism. Soc. Am. 97, S103–S114.
Vallianou, D., 1996. New evidence of earthquake destructions in Late Minoan Crete. In: Stiros, S., Jones, R.E. (Eds.), Archaeoseismology, Joint Publication by I.G.M.E. & The British School at Athens. Fitch Laboratory Occasional Paper vol. 7, pp. 153–167.
Van Dorn, W.G., 1968. Tsunamis. Contemp. Phys. 9, 145–164.
Vaněk, J., Kárník, V., Zátopek, A., Kondorskaya, N.V., et al., 1962. Standardization of magnitude scales Izvest. Acad. Sci. U.S.S.R., Geophys. Ser. 2, 153–158.
Vella, C., Demory, F., Canut, V., Dussouillez, P., Fleury, T.J., 2011. First evidence of accumulation of mega boulders on the Mediterranean rocky coast of Provence (southern France). Nat. Hazards Earth Syst. Sci. 11, 905–914.
Vida, J.M. G., Macias, J., Castro, M., de la Asuncion, M., Melini, D., Romano, F., Tonini, R., Lorito, S., Piatanesi, A., Molinari, I., 2015. Tsunami-HySEA: a GPU based model for the Italian candidate Tsunami Service Provider. Geophys. Res. Abstr., 17, EGU2015-13797-3, 2015, EGU General Assembly, 2015.
Vilibić, I., Paklar, G.B., 2006. High-frequency atmospherically-induced oscillations in the middle Adriatic coastal area. Ann. Geophys. 24, 2759–2771.
Vilibić, I., Sepić, J., Ranguelov, B., Mahović, N.S., Tinti, S., 2010. Possible atmospheric origin of the 7 May 2007 western Black Sea shelf tsunami event. J. Geophys. Res. 115, C07006, 12 pp.
Vizcaino, A., et al., 2006. Sedimentology, physical properties and ages of mass-transport deposits associated to the Marquês de Pombal Fault, Southwest Portuguese Margin. Norw. J. Geol. 86, 173–182.
Vött, A., 2007. Relative sea level changes and regional tectonic evolution of seven coastal areas in NW Greece since the mid-Holocene. Quat. Sci. Rev. 26, 894–919.
Vött, A., Bareth, G., Brückner, H., Lang, F., Sakellariou, D., Hadler, H., Ntageretzis, K., Willershäuser, T., 2011b. Olympia's harbour site Pheia (Elis, Western Peloponnese, Greece) destroyed by tsunami impact. Die Erde 142 (3), 259–288.
Vött, A., Brückner, H., May, M., Lang, F., Brockmüller, S., 2007. Late Holocene tsunami imprint at the entrance of the Ambrakian gulf (NW Greece). Méditerranée No 108.
Vött, A., Bruckner, H., May, M., Lang, F., Herd, F., Brockmüller, S., 2008. Strong tsunami impact on the Bay of Aghios Nikolaos and its environs (NW Greece) during Classical-Hellenistic times. Quat. Int. 181, 105–122.
Vött, A., Bruckner, H., May, S.M., Sakellariou, D., Nelle, O., Lang, F., Kapsimalis, V., Jahns, S., Herd, R., Handl, M., Fountoulis, I., 2009. The Lake Voulkaria (Akarnania, NW Grecce)

paleoenvironmental archive - a sediment trap for multiple tsunami impact since the mid-Holocene. Zeitschrift für Geomorphologie 53, 1–37.
Vött, A., Fischer, P., Hadler, H., Handl, M., Lang, F., Ntageretzis, K., Willershäuser, T., 2011c. Sediment burial of ancient Olympia (Peloponnese, Greece) by high-energy flood deposits-The Olympia tsunami hypothesis. 2nd INQUA-IGCP-567 Internat. Workshop on Active Tectonics, Earthquake Geology, Archaeology and Engineering, Corinth, Greece.
Vött, A., Lang, F., Brückner, H., Gaki-Papanastasiou, K., Maroukian, H., Papanastasiou, D., Giannikos, A., Hadler, H., Ntageretzis, K., Willershäuser, T., Zander, A., 2011. Sedimentological and geoarchaeological evidence of multiple tsunamigenic imprint on the Bay of Palairos-Pogonia (Akarnania, NW Greece). Quat. Int. 242, 213–239.
Vött, A., Lang, F., Brückner, H., Gaki-Papanastasiou, K., Maroukian, H., Papanastasiou, D., Giannikos, A., Hadler, H., Ntageretzis, K., Willershäuser, T., Zander, A., 2011a. Sedimentological and geoarchaeological evidence of multiple tsunamigenic imprint on the Bay of Palairos–Pogonia (Akarnania, NW Greece). Quat. Int. 242, 213–239.
Wächter, J., Babeyko, A., Fleischer, J., Häner, R., Hammitzsch, M., Kloth, A., Lendholt, M., 2012. Development of tsunami early warning systems and future challenges. Nat. Hazards Earth Syst. Sci. 12, 1923–1935.
Walder, J., Watts, Ph., Sorensen, O., Janssen, K., 2003. Tsunamis generated by subaerial mass flows. J. Geophys. Res 108 (NB5), 1–19.
Wang, X., Liu, P.L.-F., 2005. A numerical investigation of Boumerdes-Zemmouri (Algeria) earthquake and tsunami. Comput. Model. Eng. Sci. 10, 171–183.
Ward, S.N., Asphaug, E., 2003. Asteroid impact tsunami: a probabilistic hazard assessment. Icarus 145, 64–78.
Ward, S.N., Day, S., 2010. The 1963 landslide and flood at Vaiont Reservoir Italy. A tsunami ball simulation. Italian J. Geosci. 130(1), 16–26.
Watts, P., Waythomas, C., 2003. Theoretical analysis of tsunami generation by pyroclastic flows. J. Geophys. Res. 108, B12, 1–19.
Whelan, F., Kelletat, D., 2002. Geomorphic evidence and relative and absolute dating results for tsunami events on Cyprus. Sci. Tsunami Hazards 20, 3–18.
Wong, P.P., 2009. Rethinking post-tsunami integrated coastal management for Asia-Pacific. Ocean Coast. Manag. 52, 405–410.
Woodworth, P.L., Rickards, L.J., Pérez, B., 2009. A survey of European sea level infrastructure. Nat. Hazards Earth Syst. Sci. 9, 927–934.
Yalciner, A.C., Alpar, B., Altinok, Y., Özbay, I., Imamura, F., 2002. Tsunamis in the Sea of Marmara, Historical documents for the past, models for the future. Mar. Geol. 190, 445–463.
Yalçiner, A.C., Gülkan, P., Dilmen, D.I., Aytore, B., Ayca, A., Insel, I., Zaytsev, A., 2014. Evaluation of tsunami scenarios for western Peloponnese. Greece 55 (2), 485–500.
Yalciner, A.C., Kuran, U., Akyarli, A., Imamura, F., 1993. An investigation on the propagation of tsunamis in the Aegean sea by mathematical modeling. Proceedings of IUGG/IOC International Tsunami Symposium, Wakayama, Japan, August 23–27, pp. 65–75.
Yalciner, A.C., Pelinovsky, E., Talipova, T., Kurkin, A., Kozelkov, A., Zaitsev, A., 2004. Tsunamis in the Black Sea: comparison of the historical, instrumental, and numerical data. J. Geophys. Res. 109, C12023.
Yalouris, N. (1957). Dokimastikai Erevnai eis ton Kolpon Pheias Elias. Archaeoliki Ephimeris, pp. 31–43 (in Greek).
Yalouris, N. (1960). Pheia. Archaeologikon Deltion, 16 (B), 126 (in Greek).
Yanovskaya, T.B., Romanelli, F., Panza, G.F., 2003. Tsunami excitation by inland/coastal earthquakes: the Green function approach. Nat. Haz. Earth Syst. Sci. 3, 353–365.
Yelles Chaouche, A., 1991. Coastal Algerian earthquakes: a potential risk of tsunamis in western Mediterranean? Preliminary investigation. Sci. Tsunami Hazards 9, 47–54.
Yelles Chaouche, A.K., Roger, J., et al., 2009. The 1856 tsunami of Djidjelli (Eastern Algeria): seismotectonics, modelling and hazard implications for the Algerian coast. Pure Appl. Geophys. 166, 283–300.

Yiga, Y., 2003. Source process of large and significant earthquakes in 2003. Bulletin International Institute of Seismology & Earthquake Engineering 2003, 145–153.

Yokota, Y., Koketsu, K., Fujii, Y., Satake, K., Sakai, S., Shinohora, M., Kanazawa, T., 2011. Joint inversion of strong motion, teleseismic, geodetic, and tsunami datasets for the rupture process of the 2011 Tohoku earthquake. Geophys. Res. Lett. 38, L00G21.

Yolsal-Çevikbilen, S., Taymaz, T., 2012. Earthquake source parameters along the Hellenic subduction zone and numerical simulations of historical tsunamis in the Eastern Mediterranean. Tectonophysics, 536–537, 61-100.

Yolsal, S., Taymaz, T., 2010. Sensitivity analysis on relations between earthquake source rupture parameters and far-field tsunami waves: case studies in the Eastern Mediterranean Region. Turkish J. Earth Sci. 19, 313–349.

Yolsal, S., Taymaz, T., Yalçiner, A.C., 2007. Understanding tsunamis, potential source regions and tsunami-prone mechanisms in the Eastern Mediterranean, 291, Geological Society, Special Publications, London, pp. 201–230.

Yoshida, Y., Ueno, H., Muto, D., Aoki, S., 2011. Source process of the 2011 off the Pacific coast of Tohoku earthquake with the combination of teleseismic and strong motion data. Earth Planets Space 63, 565–569.

Zitellini, N., et al., 2004. Neogene through Quaternary tectonic reactivation of SW Iberian passive margin. Pure Appl. Geophys. 161, 565–587.

Zitellini, N., Gràcia, E., Matias, L., Terrinha, P., Abreu, M.A., De Alteriis, G., Henriet, J.P., Dañobeitia, J.J., Masson, D.G., Mulder, T., Ramella, R., Somoza, L., Diez, S., 2009. The Quest for the Africa-Eurasia plate boundary west of the Strait of Gibraltar. Earth Planet. Sci. Lett. 280, 13–50.

Zorrilla Aniel-Quiroga I, González, R.M., 2012. Data bases, maps and inventories for near-field tsunami sources in the North East Atlantic and Mediterranean Sea, IH Cantabria Report. Report for DG-ECHO, EU NEARTOWARN Project, July 2012.

Zorrilla Aniel-Quiroga I, González, R.M., 2013. NEARTOWARN viewer User Manual, IH Cantabria Report. Report for DG-ECHO, EU NEARTOWARN Project, 2013.

SUBJECT INDEX

E

F

G

J

K

L

M

Q

R

S

T

V

W

FSC
www.fsc.org
MIX
Papier | Fördert
gute Waldnutzung
FSC® C083411

Kolibri 360 GmbH
Ferdinand-Jühlke-Straße 7
99095 Erfurt, Deutschland
produktsicherheit@kolibri360.de